# I Learned About Flying From That

## Volume 3

Editors of Flying Magazine

## TAB Books

### Division of McGraw-Hill

New York  San Francisco  Washington, D.C.  Auckland  Bogotá
Caracas  Lisbon  London  Madrid  Mexico City  Milan
Montreal  New Delhi  San Juan  Singapore
Sydney  Tokyo  Toronto

© 1993 by **Flying Magazine**.
TAB Books is a division of McGraw-Hill, Inc.

pbk                10  11  12  13  FGR/FGR    0 9 8 7 6 5 4 3 2 1 0
hc     1  2  3  4  5   6   7   8   9  FGR/FGR  9 0 9 8 7 6 5 4 3 2 1 0

**Library of Congress Cataloging-in-Publication Data**

I learned about flying from that, volume 3 / by editors of Flying
   magazine.
         p.      cm.
      Third compilation (1988-1992) of Flying magazine's ''I learned
   about flying from that'' column. Previous volumes published under the
   titles: I learned about flying from that and More I learned about
   flying from that.
      Includes index.
      ISBN 0-8306-4281-1      ISBN 0-8306-4280-3 (pbk.)
      1. Airplanes—Piloting—Miscellanea.   2.  Air pilots—Biography.
   I.  Flying (Chicago, Ill.)
   TL710.I14   1993
   629.132′52—dc20                                            93-71
                                                              CIP

Acquisitions Editor: Jeff Worsinger
Managing Editor: Susan Wahlman
Production team: Katherine G. Brown, Director
                Linda King, Proofreading
Design team: Jaclyn J. Boone, Designer
             Brian Allison, Associate Designer
Cover design: Theresa Twigg                                    TAB3
Cover photo: Brent Blair, Harrisburg, Pa.                      4311

# Contents

# Preface

In the May 1939 issue of *Flying*, the first "I Learned About Flying From That" feature appeared. Garland Lincoln was flying a Ford Trimotor from Fairbanks into the interior of Alaska to carry fuel to other pilots searching for a missing Russian airplane. When Lincoln returned to Fairbanks, he was stuck on top of a dense fog layer with no radio and nothing but protruding mountain peaks for landmarks. After exhausting all his fuel, Lincoln managed a survivable crash landing on gooey tundra and pronounced that he learned flying into bad weather is not worth the risk, no matter how urgent the need to complete a mission.

The editors noted that "it is the belief and hope that this series of articles might save the life of some not-too-seasoned pilot. Each author will be a bona fide licensed pilot." In the 50 years since, more than 600 pilots have written of their near-misses in "I Learned About Flying From That," running up a warning flag for any who follow. Many of the mistakes were just plain stupid. Other circumstances that nearly caused disaster could never have been foreseen. No matter the cause of the problem, pilots reading "I Learned About Flying From That" are alerted to one more potential trap.

"I Learned About Flying From That" is the most widely read regular feature in *Flying* magazine. Grizzled vets and student pilots—and all levels between—learn from the experience of others. And pilots are very willing to tell others about their close calls sending in 40 to 50 harrowing stories per month for consideration by *Flying's* editors.

We've collected some more of the best "I Learned About Flying From That" stories in this third book of the same name. All stories are true, all are written by the pilot involved in the incident, and we continue to hope, as did *Flying's* editors more than 50 years ago, that pilots will learn from the mistakes of others.

J. Mac McClellan
Editor-In-Chief, *Flying*

# Introduction

We all learn from mistakes—our own and those of others. Pilots, particularly, prefer the mistakes to be someone else's. Much of learning to fly is a matter of learning through errors, and initially, our flight instructors ride along to be sure that when we make our mistakes they don't progress to the point that they become critical and develop into accidents.

On a first dual cross-country, the instructor typically lets us get off course, overfly our destination, and incorrectly tune the VOR. We're allowed to make mistakes we can learn from and build good habits upon. We graduate to solo flight when our instructors are comfortable that we've already made the critical mistakes when they've been with us and have only smaller, non-life-threatening errors still to learn from.

*Flying* magazine's monthly column, "I Learned About Flying From That," lets us all learn from the mistakes made by other pilots. The accounts are submitted by all kinds of pilots—instructors, students, airline, and military—each of them still learning, but willing to share their often-embarrassing experiences so others can learn without having to make the same errors.

This collection of ILAFFT columns that appeared during the seven years beginning in March 1986 once again shows that, although the mistakes people make are unique and individual, they often have elements in common. These are accounts of people who erred and, through luck or skill—and often both— were allowed to come back to fly again, and ideally, never make the same mistake again. Some are funny, some frightening, some amusing. Read them and weep, smile, shudder, or cringe, but they're presented here for you to learn more about flying from them. Remember, it's much easier on you and your airplane if the mistakes you learn from are someone else's.

Tom Benenson
Senior Editor, *Flying*

# 1

# Learning to Show Proper Respect

Well before man lifted himself off the surface of the earth, thunderstorms already demanded and were granted a great deal of respect. From our first lessons, we're all taught to steer clear and give storms a wide berth. When a thunderstorm really gets going, it can wield more force than most aircraft can fly against. Nevertheless, planes and thunderstorms do end up sharing the same airspace. And it occurs much more frequently than is healthy for either the planes or their pilots.

Weather has always played an important role in accident scenarios, and it's thunderstorms that most often get to take the dramatic villains' parts. They can be very convincing as the bad guys. The following accounts detail experiences of pilots who have confronted thunderstorms—sometimes intentionally (and foolishly) and sometimes inadvertently.

No one wins a bout with a boomer—a pilot can occasionally run away and get to fly another day. All of the chroniclers confessed to learning never to fool with Mother Nature.

## Radar Can Lie
### by Dave Southworth

My first officer greeted me with a pleasant smile outside the Metroliner II. The day was starting out like any other summer day: the weather report at our destination airport called for clear skies, 85°, light winds and a chance of cumulus buildups in the late afternoon. In no time we had the airplane ready to go with eight excited passengers strapped in behind us. The

1

turbines screamed to life, and we departed the ramp on time, leaving behind only the lingering smell of jet-A.

Leveling off at 20,000 feet we settled back and prepared ourselves for the first leg of our eight-leg day. No problem, I thought. Smooth, blue sky; what a life!

As I flew along, I saw cloud towers starting to build on the horizon. I mentioned that it might get bumpy later on. With a passing smile my partner nodded his head, switched on the radar and carefully adjusted the antenna.

We were coming up on the first of two uncontrolled airports that were 80 miles apart, and our destination was the second one. Between IFR arrivals, departures and VFR advisories, center was busy that afternoon. The one man working the frequency had his hands full.

"Descend pilot's discretion to one four thousand." That was for us.

"Descent check," I said. Pulling the power back I started the Metro's smooth 200-knot descent out of 20,000 for 14,000 feet.

Level at 14,000, we could see we were above the bases of a few isolated buildups in the area. We could also see a kind of dense haze layer. Part of it was weather building, part of it smoke from a nearby forest fire.

"Check the radar," I said.

There was no weather return on the screen, but the radar would paint the ground, and it seemed to be working properly. We couldn't see around the haze, and I knew that trying to circle it would be an all-day project.

"Ask center if they show any weather ahead and ask for 12,000. Maybe we can go under this." I had a feeling that it might get bumpy in there, and I didn't want to upset the passengers.

Our altitude request was instantly denied because of conflicting traffic. No weather was noted on center radar.

"Let's tell the folks to tighten up their seat belts. Also mention to center that we want 12,000 as soon as possible." I tightened my belt, turned on the continuous ignition, windshield heat, pitot heat, intake and prop heat, just as the book says to do when you're flying in clouds. We entered the clouds about 200 feet above the bases. For the first few minutes it was smooth.

Then it started, fight turbulence, and the cockpit darkened. I had to turn on the instrument panel lights. A chill ran down my back; this was not good. The turbulence became more intense. I glanced at the radar: nothing but a clean screen.

I still didn't like it. It *felt* bad. From the look in my partner's eyes, I could tell our feelings were the same.

"Let's get a 180 out of here," I said. The first officer's finger was already on the button, and he asked for a 180-degree turn. Due to frequency congestion, his request became static gibberish. "Try again if you can," I said, as I pulled the power back to 194 knots, which was rough-air maneuvering speed.

The airplane rocked violently, nearly jerking the controls from my hands. The nose went up, and we entered a 2,000-fpm climb. I pulled the power back to flight idle, pushed the nose down and tried to maintain our assigned altitude. It was no use; with no power on, the airspeed still shot up to 240 knots, just below redline. The airplane kept climbing at 2,000 fpm.

Then came the hail. When it hit the airplane my heart stopped. It was no average hail. Estimates later showed some stones to be more than three inches in diameter. The airplane was all over the sky, 60- to 90-degree banks each way. The hail hit our stall avoidance system, destroying the angle-of-attack vane on the right wing. The stall horns went off, and the needle on the dash indicator slammed into the red.

"SAS clutch!" I had to yell into the intercom to be heard over the deafening noise. The first officer reached instantly for the switch, shutting off the system. At the same time I fumbled for the computer circuit breaker.

Visions of the windshield giving way streaked through my mind. Flying glass and depressurization. Dear God, I hope they hold. The airplane yawed sharply left. I glanced down to see the instruments on the left engine heading for zero. We had a flameout.

"We lost the left one." My words sounded as if they'd come from someone else. I shot a quick look at my partner. I could see the terror in his eyes. I wondered if he could see it in mine.

Thank God for systems that work. The auto continuous ignition refired the engine at 90-percent rpm, as advertised. We once again had two engines. The airplane was coming apart, and I had no idea how long this weather was going to hold. Should I turn around or should I go on? I knew that all the books say, "Go on, it won't last," but at this rate the airplane wasn't going to last either. I started a left 180.

Sometimes moments can seem like hours. The entire storm lasted about two minutes. As I turned to the left, sunlight broke through our right windshield. I turned back to the right and came out into clear blue sky.

"Cleared for the visual approach," center barked in our headsets. "Roger," was the weak reply from my partner. The landing was normal. Aircraft damage was substantial: numerous dents in the leading edges of the wings, the fuselage, around the nose area, and the horizontal and vertical stabilizers. Several dents on the wings were three to four inches in diameter and two to three inches deep, popping rivets and separating the skin. Both wingtips were shredded. The rotating beacon, along with the entire top of the vertical stabilizer, was missing, presumed dead. There was virtually nothing left of the radome on the nose but pulverized fiberglass.

What did I learn? I learned that the radar in most cases is a valuable tool, but it can't be solely relied upon. The radar equipment was checked upon our return home and given a clean bill of health. Why it didn't show the storm remains a mystery. Still, the pilots are the most important equipment in an airplane. I should have followed my instincts and turned around when we were in contact with center for a lower altitude. We could have avoided the whole incident. As well as learning not to trust some of the airplane's equipment, I learned to trust other systems with our very lives. If I hadn't turned on all of our anti-ice systems as well as the auto continuous ignition system, the result of the storm would have been very different, possibly even tragic.

Finally, I learned to have a great deal more respect for Mother Nature. The centers of those large cotton-like clouds floating peacefully in the sky house more energy than anything man has yet built.

All of the passengers made a trip around the airplane before going into the terminal. "Were you scared?" one older gentleman asked pleasantly. I just smiled at him.

*The Metroliner pilots knew that thunderstorms should be avoided, but their radar didn't provide any clues that a real granddaddy was out there waiting for them.*

*In the next incident, a checkout in a Cessna Turbo 210 was given a new twist when thunderstorms sprinted ahead of their forecast pace and arrived on the scene earlier than expected.*

# Twister

### *by Ray Klaus*

The day started out like any other summer day in August: sunshine, scattered clouds, hot humid, hazy air with light winds and a chance of thunderstorms in the late afternoon. Around midmorning, I received a telephone call from a customer based with the FBO for whom I worked at the Chicago-Aurora Airport. The pilot owned a Bellanca Viking, but wanted a checkout in our six-seat Cessna Turbo 210 to take some business associates to Louisville, Kentucky, the next day.

We arranged to get together after lunch. A check of the weather revealed level five radar returns popping up northwest of Rockford, Illinois, about 90 miles away. There was a lot of energy in the sky that day, and the forecast called for thunderstorms moving our way.

It was about two p.m. when we finished reviewing the performance charts, aircraft systems and operations of the 210 and took off. In flight, we did the usual airwork and aircraft familiarization associated with a complex aircraft checkout and returned to the airport for takeoffs and landings.

The sky was darkening to the north with the approaching storms, but they appeared far enough away not to pose any immediate threat. To give us a bit of margin and an escape route, if needed, I requested a left-hand pattern for Runway 27 instead of the standard right-hand pattern.

On downwind, the tower broadcast "Winds calm." We did a stop-and-go on the 6,500-foot runway: smooth as silk. Next time around, same thing. "One more time and we'll be finished," I said. Little did I know that we might really be finished.

The third time on downwind, the tower was still reporting winds calm. As we turned base, the winds perked up to 15, gusting to 20 knots. Still nothing to get alarmed about, but the sky was getting darker. Turning final, I commented to the pilot I was checking out that I thought the rain would hit before we got to the ramp and that we'd probably have to wait to be rescued by someone with an umbrella to keep from drowning in the deluge.

Halfway down final approach, the sky turned black—a black I had never seen before. No reflected light. Nothing. It was akin to flying into a pool of India ink. The runway lights popped on immediately. I commented to the tower, "Nice display of lights." "We like 'em too, Ray," they responded. There was still no hint of the lurking danger, the air was relatively smooth and we couldn't see what was to befall us in the darkness.

In the flare for touchdown the 210 was suddenly grasped in the jaws of a monster and pitched, rolled and yawed beyond belief. The situation was, to say the least, critical. My singular thought was of survival—"Keep the airplane flying right side up and get out of here!" I grabbed the controls with "I've got it," tightened my grip to keep them from being wrenched out of my grasp, and called for the other pilot to add max power and raise the flaps to approach position. We were climbing at a fantastic rate with the gear coming up.

The airplane was being badly battered about as we rode the updraft—"shaking like a pair of dice in the hands of the devil," as the old saying goes. I thought the Cessna would break apart. So I called for power reduction, slowed to maneuvering speed, lowered approach flaps and gear. Within a blink of an eye, we were being bashed back to earth. I called for max power again, with approach flaps and gear up. Nothing arrested our descent. The granules of sand in the concrete runway zoomed to huge dimension before our eyes.

Impact appeared imminent. However, just before being atomized into oblivion, we seemed to cushion in ground effect and began climbing out. Since the first encounter, we had not made much forward progress—just up and down like riding an elevator operated by a madman.

At midfield, with no more than a couple of hundred feet between us and the ground, I cautiously edged away from the storm, turning downwind toward the open sky. It was the only good place to go.

Suddenly, the airplane rolled near-inverted. I could see the numbers for Runway 18 in the top of the windscreen. I pressed the yoke forward to keep from split-essing into the ground while using aileron and rudder to roll the airplane right-side-up.

The people on the ground said the entire scenario was a scary sight. They could clearly see the storm approaching with twin vortices sucking up top-soil and vegetation from the farmland upwind of the airport. All available manpower was mustered to hangar and secure airplanes. Pilots in a Citation, holding short for our arrival, observed our wild flight with disbelief. As the airplane rolled nearly inverted, it seemed destined to crash into a large storage hangar complex filled with bizjets. As we were blown away from the scene, the airplane pitched, yawed and rolled unlike anything they had ever seen. As a matter of fact, when we were rolled inverted and I was just managing to get it right-side-up over the airport, the tower tersely reported, "Winds exceeding 60 knots, window glass cracked. We're evacuating the tower. Good luck, Ray!"

We headed for the blue sky, fighting turbulence at maneuvering speed, and landed 50 miles away at Ottawa, Illinois. A call was made to flight service with the query, "What hit Aurora Airport?" "Some heavy-duty thunderstorms were reported passing through, but the area will be clear in about 30 minutes," they reported. That was all it was?

At Ottawa, a careful visual inspection of the aircraft revealed no physical damage—not a rivet out of place; no dents or cracks. This conclusion was later confirmed by shop inspection. The Cessna T210 proved to be a sturdy bird. It no doubt helped us survive the experience.

The other pilot's calm, immediate response to my calls for power, flaps and gear adjustments made up the balance in our survival. I had my hands full with the flight controls. It was definitely a two-person job.

Upon landing back at Aurora Airport and taxiing to the ramp, I noticed a Cessna Skymaster over on its back, torn from its chained tie-down. Metal hangar siding was twisted and peeled back from hangars adjoining the ramp. Farther downfield, abeam the spot of our initial encounter, gave witness to the full destructive fury of the storm. Two rows of T-hangars had been completely raised from their foundations and thrown several hundred yards downwind. A beautiful Cessna 310R was smashed upside down. The tie-down rings attached to the chains on a Piper Seneca were pulled right out of the wings as the aircraft had been lifted, spun around and smashed back on the ramp into a crumpled mess. In all, some 32 aircraft were destroyed that day at Aurora Airport plus extensive property damage.

Meteorologists had issued a severe thunderstorm watch for the region that afternoon, but had not foreseen a tornado. Instead of dissipating as expected as it moved to the southeast, the storm grew stronger. Then at its core, amid powerful counterclockwise winds and updrafts pushing 60,000 feet above the earth, the tornado was born.

This one had widened rapidly, spawning a cluster of smaller twisters. Inside the Hydra-headed monster, winds built to an estimated 275 to 300 mph. By the time the tornado crossed downline communities, it was so powerful that it stripped bark from trees and gouged earth from the fields.

North of Rockford, Illinois, where it started, meteorologists estimated the storm's forward speed was 30 to 35 mph. By the time it passed the Aurora Airport, the storm was moving up to 60 mph.

It was not a textbook twister. It did not follow the patterns, did not have the looks and did not until the last few minutes, act like what it was—one of the most devastating storms in northern Illinois history.

What did I learn from this experience? First of all, storms can move much faster than normally anticipated. I thought we had two to three hours' time before it hit. Instead, it overran the airport in a little over an hour.

Tornadoes do not necessarily move only from the southwest to the northeast. This one came from the northwest to the southeast.

Finally, I learned to have a great deal more respect for Mother Nature. The trick in dealing with any kind of severe weather is to keep as much distance between it and you as possible.

✈ ✈ ✈

*The pilots in the 210 survived their encounter with one of nature's deadliest forces, because, after initially underestimating the speed of the storm, they made all the correct decisions and worked together as a crew. And they were lucky.*

*The Corsair pilot who confessed to the next encounter had three opportunities to make the wise and prudent choice. But he didn't. He continued with his approach almost until it was too late to break it off. He, too, was lucky. The approach could have broken him off.*

# A-7 Coarse Air

## *by Boehmer Jon Gorr*

I was flying a single-seat A-7D Corsair II tactical fighter on a two-ship training flight from Greater Pittsburgh International Airport. I was the flight leader and my wingman was a pilot for a major airline but new to the A-7D. The weather forecast for that summer afternoon included the possibility of severe thunderstorms. Although we had a color weather radar display at the base, the supervisor of flying for the day was not updating it, and I failed to ask for an update. Without getting the latest available weather information, I walked out the door to the airplane.

After we took off and penetrated some broken decks, we entered clear air and saw a massive line of thunderstorms, brilliant and beautiful at altitude. Even though I knew the westward flow of the storms would carry them near the field, I failed to call back a pirep. After we completed our mission and before returning to Pittsburgh, I worked with Indianapolis Center to split up our two-ship flight. I didn't want to be negotiating thunderstorms in formation. As standard procedure, I sent the inexperienced wingman in first so that if he had a malfunction requiring a chase plane I would be somewhere behind him. That was the only smart thing I did that afternoon.

Pittsburgh Approach vectored me clockwise around the field the long way instead of putting me on the shortest route for an ILS approach to the west. As I flew in the clouds, I did some knob-turning on the radar and saw that the line of sharply defined weather was near the field. At that point I should have requested holding instructions. I had enough fuel to hold for half an hour and go to alternates at either Columbus, Ohio, or Harrisburg, Pennsylvania. When the Pittsburgh controller asked me if I wanted to hold or try an ILS, he also said that precip was moving rapidly toward the final approach course. During the entire episode, Pittsburgh Approach did a superb job. Once again, I had the opportunity to hold, but I asked to be vectored to the localizer. Before turning base for the ILS, I was briefly in clear air and saw a major storm with lightning between the outer marker and the field. I thought I would pop through the rain and into clear air while still a mile from touchdown. After I was established on the localizer, though, I heard my wingman request to break off his approach. The controller asked if I would still like to continue. After telling him that I would, I set my radar altimeter's low-altitude warning at 300 feet and descended toward the black curtain that stood tall in front of me.

I thought I would be jostled a little and then break into the clear for landing. When I hit the curtain, however, I entered nightlike darkness, and as the water hit the windshield it sounded like rifle cracks. Lightning flashes turned night into whiteout, and I experienced difficulty reading the instruments. I began to think that I was a fool for attempting to rush the approach. Suddenly a brilliant flash left me blinded and I rose violently against my shoulder straps as the aircraft bolted downward. I had never felt such severe vertical acceleration. While I was in this elevator shaft, the 300-foot altitude warning sounded in my helmet. I pushed the throttle to the stop and squeezed backward on the stick. My vision started to return, but at first I saw only green darts and tadpoles shooting around. Then I realized that I was in a nose-high attitude, the aircraft was still heaving from turbulence and a waterfall was cascading onto the windshield. I asked Pittsburgh if I could break off the approach, and then I did probably the worst job in aviation history of complying with missed-approach instructions.

I continued to bounce in turbulence, heavy rain and lightning as I overcontrolled the airplane en route to the holding fix. My first attempt to enter the pattern was foiled when I set the wrong navaid frequency.

After one turn in the holding pattern, approach vectored me to clear air over the field and set me up for a right base to the same runway on which I had just attempted to land. The storm had departed quickly to the east.

I had heard about flash blindness but had never experienced it. Then I remembered the Delta L-1011 at Dallas/Fort Worth and wondered why it's so difficult for us to learn simple lessons about thunderstorms.

*Why, he wonders, is it so difficult for us to learn simple lessons about thunderstorms. Why, indeed! In the next example of thunderstorm penetration, the pilot learned his lesson the hard way when he decided to go through what he thought was a hole in a line of thunderstorms.*

# From the Belly of the Beast
## by A. J. Thieblot

We were heading back toward Baltimore after a day's fishing on the Outer Banks of North Carolina when we hit the first turbulence. The accelerometer in my Piper Navajo's flight director bleated in protest and tripped out the autopilot. I smiled my best rueful smile over my shoulder to reassure the passengers, pushed the buttons to reset things and went charging into what the radar suggested might be a hole in the line of thunderstorms that hung over the North Carolina-Virginia border. I found out fast that there was definitely no hole, and that we were in deep trouble.

Something was pushing the left wing up much harder than I could pull it back down, forcing the airplane hard right, and the little wings in the artificial-horizon indicator fluttered like a hummingbird's.

I froze. For a while, all I could do was stare at the artificial horizon and wonder what was happening. Then the hours of training paid off (or maybe I just didn't have a better idea), and I let go of the controls, pulled the throttle back, and let the airplane sort itself out as best it could. Even so, the left wing kept rising until it was almost vertical.

Then we hit another bump. The accelerometer blared and the autopilot tripped off again. Suddenly we came barreling out of the bottom of the cloud going downhill at 4,000 feet a minute.

The gods who protect fools were on the job that evening, so I didn't pull the wings off or furrow the North Carolina soil. I struggled back toward the southeast in an attempt to get in front of the advancing storm. I made an NDB approach at the first airport recommended by the 911 feature of my loran, which turned out to be Plymouth, North Carolina. I landed there a few minutes ahead of the storm's roll cloud.

Once we were all safe on the ground, I discovered that 120 pounds of tuna stuck to the cabin ceiling makes an awful mess.

The inside of the airplane smelled terrible for weeks, but that was far from the worst result of our brief encounter with the thunderstorm. During those few minutes of total confusion before I pulled the throttles, the engines must have run on the hot side of peak for just long enough to turn the exhaust valves into some fancy art deco stemware. One engine needed a major overhaul, and the other took three new cylinders before we could fly again.

I'd had good, groundbased weather reports from the TV weather channel, an excellent color radar and to top it off a WX-10 Stormscope, all working, and all doing their best to tell me about the storms. But I was tired and sunburned and wanted to get home, so I ignored the warnings and was lucky to survive.

The other problem was with the autopilot. The roll-control servo motor failed as I entered the storm and began commanding only right-hand turns. I had no clue that it had failed, not even a warning light.

Fortunately the turbulence was severe enough to knock the autopilot off-line again before the airplane rolled inverted.

My airplane is equipped with some complex equipment that is not all that common in smaller general aviation airplanes, but the same failure could happen when flying with even the most primitive wing-leveler. Should that happen in a thunderstorm, you may find yourself trying to sort out what's going on at a time when all of your skill is needed just to keep right side up.

In thunderstorms and turbulence it's probably best to turn the wing-leveler on and the altitude-hold off, as the instruction book says.

But if you suddenly find yourself careening out of control, kill your autopilot and try to remember how airplanes used to work before they got fancy.

✈ ✈ ✈

*"But I was tired and sunburned and wanted to get home, so I ignored the warnings and was lucky to survive." We pay for radar and Stormscopes to warn us about the presence of thunderstorms and then we choose to ignore their messages. Not smart.*

*The pilot of a Piper Seneca in the next tale knew where the storm was when he was vectored toward it by air traffic controllers. Nevertheless, he still thought he had a sufficient safety margin.*

## Squeeze Play
### by Phil Steeves

The first recorded instance of a pilot caught in a navigational dilemma was Odysseus, who had to thread his legendary ship between Scylla (the monster on the rocks) and Charybdis (the whirlpool). On this hot July day, it seemed that nature and ATC conspired to spring their version of this classic trap on me.

We were returning to Boston in our Piper Seneca after a weekend in Nova Scotia. We stopped in Bangor, Maine, to clear customs and get the latest weather update, which included typical summer thunderstorms moving eastward across southern New Hampshire.

With both radar and Stormscope on board, we were well equipped to handle it. We flew around a few heavier showers, and through a few lighter ones, all of which painted on the radar but not the Stormscope. Throughout all our maneuvering the ride remained smooth.

But then the Stormscope began indicating a distant area of activity at 12 o'clock. As we came within 100 miles, it showed clearly on the radar as an intense thunderstorm with well-defined margins and a sharp gradient—the typical isolated summer storm.

We were in instrument conditions at 6,000, so this was an embedded storm. Tops were reported at Flight Level 410, with more storms coming in from the west. It was clearly one to avoid.

As we neared the storm, the Stormscope lit up. Even on the 25-mile range, the dots were being replaced so quickly that the scope had that distinctive churning effect that practically shouts, "Keep out!" I established a deviation course that would give us a 10-mile separation from the storm.

As the threatening cloud mass began to pass off to our right, the radar image slipped off the screen, since it has only a 120-degree forward view. But the Stormscope continued to display the storm well.

I started to swing gradually around to the southwest, keeping the storm at three o'clock, maintaining what I was sure was a 10-mile distance. We were still in the clouds, but the sky was bright, and there were only a few bumps.

Home was only 15 minutes away when suddenly Pease Approach said, "Seneca 67X, turn immediately right 30 degrees." On instinct, I started the turn, but abruptly caught myself.

I had not heard Pease Approach working any traffic on the frequency in my vicinity. But since Pease was an Air Force base, they might well be working military traffic on UHF rather than VHF, and I would not necessarily hear their transmissions. I suddenly had a vivid mental image of a flight of F-111s or a KC-135 bearing down on me from my left (there was my Scylla), and ATC was telling me to turn immediately right, where lay an airborne maelstrom as fearful as Odysseus' Charybdis.

Whenever you get an immediate turn from ATC, your response should be *turn first* and ask later. I hesitated for, say, two-tenths of a second, and said, "A turn to the right will put me into that storm; how about a vector left?" The controller responded in an anxious voice, high-pitched and very insistent, "Unable; turn immediately right 30 degrees."

So I did. I reasoned that a midair was sure to be fatal. I thought I could get by with a right turn for a minute or two, followed by a left turn away from the storm as soon as the controller said I was clear of traffic.

I rolled out, double-checked that I was at my assigned altitude, and tried to reorient the radar. I was just about to ask for a turn back to the left when the sky abruptly turned dark green; I knew I was in for it.

There was no way I was going to tackle that storm, KC-135 or not. The airplane began buffeting furiously. I disengaged the autopilot, but not before it had trimmed nose-down against the strong updrafts, leaving me with startlingly heavy stick forces.

The airspeed was well into the yellow, even though I had already extended the speed brakes. The throttles had to come so far back that the gear-warning horn was sounding, which together with the autopilot disengage alert added to the general melee. In the middle of this, approach called, asking me to squawk ident. I had my hands full, and bouncing around like that I couldn't get a hand on the ident button.

But I did manage to transmit, "You've vectored me into a cell, and I'm fighting like hell to keep it under control."

Now, there is a placard on my instrument panel with a quote from the Roman philosopher Seneca that reads: *"Bunvernator qui boc potuit dicere 'Neptune, numquam banc navem nisi rectam,' arti satis fecit."* That means, "The pilot who is able to say, 'Neptune, you will never sink this ship except on an even keel,' has satisfied the requirements of his art." Loosely translated: fly the airplane first. So I did.

Approach called back 30 seconds later, asking again for an ident. I told him to stand by, as much to get him off my back as to reassure him I was still flying. I had already started a gradual left turn when the excitement had hit; we exited the cell after no more than two or three minutes, having probably just flirted with the edge.

That old rule that says not to turn around while in a thunderstorm doesn't apply if you have a good picture of the storm's location, size and shape. Once we were stabilized I notified approach of our situation and told him we'd discuss it on the ground. The rest of the flight was routine, and I was able to phone approach after landing and found out that my traffic had been a Cessna, IFR at 4,000, without a Mode C transponder, whom approach was working on a different frequency. I had been overtaking him, and as our targets came closer, approach asked him to verify his altitude.

That's when he reported that he had encountered a severe updraft and was at 5,500 feet and climbing. Our targets actually merged momentarily, so I assume that we came quite close to colliding.

So what did I learn from all this? That the final authority rests with the pilot? No; I already knew that. And it was no news to me that ATC directions should always be accepted with a healthy dose of skepticism. There are enough accounts on record where ATC vectors have sent unaware pilots into "cumulogranitus." But my 10-mile separation from thunderstorms rule of thumb may not be adequate. From now on I'll give those killers wider berth, to allow more room for that one-in-a-million possibility, and to prevent getting trapped between the rocks and the whirlpool again.

# 2

# Flying with Lady Luck in the Right Seat

When the dust has settled and the investigators have turned in their reports, there's often only one explanation for the survival of a pilot and his passengers. Some call it luck, some fate, some a miracle. It's the only way to explain why, with everything stacked against a successful outcome, everyone got out and was given a second chance.

Pilots are not taught to rely on luck. Emergencies are anticipated, simulated, and practiced until the proper response is so ingrained that an emergency is nothing more than another routine maneuver. Of course, that's assuming that every conceivable combination of failures can be anticipated. Unfortunately, when something does go wrong, it's often unique and all the training won't have prepared the pilot to handle it.

Ernie Gann wrote that God was his copilot. God or Lady Luck, someone or something (whatever your beliefs) was riding along when these pilots got into trouble—and then got out again.

✈ ✈ ✈

## Bushwhacked by Canoe
### by David McAskill

I was a pilot for a small air service in the thick of the northern Ontario bush when a routine flight turned into a scary and bizarre fiasco. The flight happened on a calm, sunny afternoon in mid-July on board a 1968 Cessna 180. The airplane was equipped with oversized floats capable of displacing about 300 more pounds than the maximum gross weight of the aircraft. My trip that day was to an outpost on the shore of a small lake

60 miles northeast of our base. I was to deliver a brand-new canoe to the lake and take back an older canoe in need of repair.

Something the size of a canoe must be attached externally to the side of a Cessna 180. Although this was the first time I had ever flown with an external load, I had discussed the handling techniques with some of the more experienced pilots in the area. The basic rules were simple: use 10 degrees of flaps, don't exceed 80 knots, keep all turns shallow and make sure the load is securely tied to the aircraft.

With all this knowledge lodged firmly in my mind, I felt confident about the trip. One of the outpost guides, an Ojibwa Indian named Stanley, offered to come along and lend me a hand. We had lunch and then walked toward the dock to set off. To my surprise the 180 was sitting at the dock with the canoe already attached. My boss had roped the canoe to the aircraft while we were eating. Despite his experience (he was also a pilot), I was no fool and thoroughly inspected each knot before making the trip.

Everything looking good, Stanley and I took off toward the northeast, landing safely about 50 minutes later. We unraveled all of the ropes holding the canoe in place and carried it to shore. The old canoe sat underneath the elevated outpost; it appeared to be slightly heavier and larger than the newer one.

At this point I picked up one of the unraveled ropes lying on the dock, looked at the canoe and thought to myself, What do I do now? In all of my 22 years of existence I had never tied a canoe to an airplane. In fact, I hadn't tied much of anything onto anything. I stood there for a few seconds as Stanley stared at me confidently. I decided that I was not about to shatter Stanley's image of me. After all, those famous bush pilots of the past probably didn't have anybody around to tell them how to do things like this; they apparently just improvised, so that's what I did.

When I felt I had tied the canoe firmly in place, we were ready to depart. I set the flaps at 10 degrees, taxied out and departed into the wind, the 230-bp Continental engine pulling us smoothly up to 2,500 feet.

A few minutes after leveling off I looked out the right side and saw the canoe moving as if it was pivoting about its center. I briefly recalled the story told to me about the canoe that accidentally released from an aircraft in flight, striking the horizontal stabilizer and forcing the airplane straight into the ground. It was definitely time to find a lake, set down and secure the wobbling canoe.

Underneath us was a cluster of small lakes, but even from 2,500 feet I could see that they were clogged with reefs and floating logs—I wasn't that desperate to land. There was a lake familiar to me about seven miles to the west of our present position, so I pointed the airplane that way and started a shallow descent.

In my haste I had let the airspeed build up to 105 knots. I quickly corrected by raising the nose. Seconds later Stanley nudged me and pointed toward the canoe. What I saw was our canoe fighting against the strain of the ropes and creeping slowly rearward. As if this wasn't enough torture, one of the ropes worked free and started hammering against the fuselage.

As we passed through 2,000 feet the real fun began. Without any warning the 180 decelerated and banked violently nose-down to the right. I struggled to keep the wings level, and as I started pulling back slowly on the control column the stall warning blared. What was going on? Even with the prop and power levers full forward, I was unable to hold altitude. The VSI indicated a descent of 700 fpm, then dropped swiftly to 1,000 fpm, then erratically bounced up to 400 fpm.

I was sure that the canoe had slipped and struck the horizontal stabilizer. I looked behind me to discover that I was wrong. There was no damage. What I saw instead was the canoe still attached to the aircraft by one rope, fluttering playfully about six feet behind the tail.

I sank into a state of controlled panic. Here I was flying an aircraft unable to hold altitude under full power with a canoe dancing in the wind several feet behind. While I sat dazed, Stanley quickly pulled out his fillet knife. At first I thought he was preparing to kill me for putting him in this uncomfortable spot. But I was wrong. Without hesitation he slid back his seat and cracked open the right door. Fighting against the blast of air, he worked half of his torso outside while I kept my right arm wrapped around his legs and used my left hand to fly the airplane. He tried, but he couldn't reach back far enough to cut us free. At this point I felt that it was all over.

We were about 500 feet agl, descending, when Stanley got back in the cockpit. All I could see from our altitude were trees. The lake I was heading for had disappeared. The VSI eased to 300 fpm, but with no place to land and a few hundred feet of altitude left it didn't seem to make much difference. I decided to cut power and fuel just before hitting the trees and told myself that the floats would absorb the initial impact.

At less than 100 feet above the trees a river flashed into view directly ahead of us. We were almost perfectly positioned to land. At about 20 feet I held my breath and hauled back on the yoke. The landing was like going through a car wash at two Gs. I was completely disoriented. The engine quit and the airplane came to a quick stop as it bobbed on the surface after completing what must have been the world's shortest floatplane landing. The nightmare had ended. I could hardly believe that the airplane had not flipped over. I was certain that the oversized floats took most of the credit for keeping us upright. Unbelievably, the canoe was still hanging onto the 180 by that rope but in considerably worse shape than when we first saw it back at the outpost. The 180 started drifting to shore, so I jumped into the shallow water and beached it. Stanley and I sat on a large rock near the airplane and silently contemplated our fate.

I learned from this experience that my strong image of what a bush pilot is expected to be got in the way of sound judgment. I believed that it was my responsibility to tie that canoe to the 180 without any knowledge of an accepted technique. As it turns out, there is a complex strapping technique, which I have since learned.

The next time you step up to any aircraft, fight off the "go for it" feeling and learn the ropes before pressing on.

✈ ✈ ✈

*Lashing a canoe to an airplane has a historical precedent. The Wright brothers flew one of their planes with a canoe tied underneath it in case of a water landing in New York harbor. Can you imagine looking back and seeing a canoe chasing you? Quick, paddle as fast as you can! Amusing as it is now, it certainly wasn't at the time.*

*Pilots know better than to let it all hang out—but sometimes they forget or have no choice. Interrupted during his preflight, the pilot of a Piper Comanche almost bought the farm instead of just plowing a furrow.*

# Have a Good Trip
## *by Don Ostergard*

Southern Alberta is a terrific place to fly most days of the year, and August 11, 1984, was one of those times. A sky that seemingly reached forever, the sight of the Dinosaur Valley badlands cutting through the farms and ranches, the Rockies in the background—all were compelling reasons to travel via Comanche rather than Pontiac the 60 nm from our farm near Drumheller to the family picnic in Calgary.

My wife, son and I preflighted our airplane and had just pulled it out of the hangar when the phone rang. Being a slave to the darn thing, I ran to answer it. A neighbor wanted to borrow something. I walked back to the airplane, annoyed at myself for dropping everything and dashing madly across the yard to answer a phone that was ringing on a Saturday afternoon. We got in and fired up, taxied to the far end of our airstrip, went through the checklist and started our takeoff roll.

Just as I rotated we experienced a most unforgettable bang and shudder. Had our puppy followed us out to the strip? No. The Comanche had just pole-vaulted over its towbar. We had taxied 200 feet and made two 90-degree turns, then back taxied 2,500 feet, made a 180 and accelerated to 70 knots—all on grass—with a seven-foot-long towbar sliding in front of the airplane.

Now what? Everything seemed normal—no strange noises, control pressures or vibrations. It seemed like a good idea to keep both the landing gear and the speed down. A look at the mirrors on the tip tanks (installed to help guard against that other dreaded inevitability) showed the nose gear to be properly pointed downward, but how much detail do you get from a three-inch convex mirror?

Since there was no compelling reason to land right away, I ruled out a return to our farm strip. (If we were going to crash we certainly didn't want it to happen out in the country with no one to see it.) So we called up nearby Drumheller Airport's unicom to advise them of our predicament and to ask if someone would come out and have a look at the gear while we made a couple of low passes. If the gear looked okay we would land there. If not, we would fly to Calgary, where 12,675 feet of asphalt and a great array of emergency equipment would welcome us. (The fellows at the Drumheller airport find themselves short of breath after pushing their two-wheel fire cart the length of the runway.)

It happened that one of the local spray pilots had just taken off in his AgTruck; he offered to fly alongside and have a look at the nose gear. He couldn't see anything wrong, so we prepared to land. We cleared everything off the hat shelf, and to bring the CG farther aft, my wife climbed into the back seat. For a fleeting moment I found myself wishing she was obese.

I managed a perfectly smooth landing (a not unremarkable event in itself) and rolled out without the nose gear collapsing. Safely parked on the ramp we got out to look at the damage. The nose-gear assembly and firewall were unscathed. In fact, everything looked fine until we got to the tail section. The tow-bar had struck the starboard side of the stabilator, forcing it backward slightly and causing its supporting bulkhead to be partially torn from the fuselage skin where it attaches. Good thing we had kept the speed down. Any flutter at all would have ripped out the few remaining rivets, and the stabilator would have been lost. So would we.

The Comanche is flying again. The paint match is pretty good, and most people never notice the repairs. As a result of my experience, if I am interrupted again during either my walk-around or my checklist, I start over. If the phone rings, I am still fool enough to run to answer it, but the new towbar I have built has a big spike pointing down and forward at 45 degrees, just in case.

✈ ✈ ✈

*It's not smart to tow a towbar, but at least it doesn't add as much drag as a dangling canoe. In the next account, a commuter crew on a night flight in a Beech 99 found themselves being pulled down—even at full power—right toward their own accident site.*

# Dragging an Engine
## by Henrik Wynne

I was flying as first officer on a Beech 99 commuter turboprop from Inuvik in the Canadian Western Arctic to Fairbanks, Alaska. We were empty. We had filed IFR for this night flight, but the weather en route was reported to be good. We leveled off at 8,000 feet on top, settling in for the two-hour flight, enjoying the full moon and the smooth ride in the cold still air.

After 30 minutes in the air we passed the boundary of the overcast below us, leaving us to enjoy the Richardson Mountains bathed in moonlight and the lights from the small Indian community of Old Crow as we passed overhead.

A few minutes before reaching the Canada/Alaska border at 141 degrees west, we encountered some light chop. When checking in with Anchorage Center, at the border, we requested 6,000, hoping to get out of the bumps. This put us back in smooth air and everything was quiet again.

Two minutes later there was a very loud *boom* from the right engine. When I looked out I saw sparks and flames streaming out of the exhaust stack of the good old Pratt & Whitney PT6 engine, lighting it and the fuselage up. The airplane yawed violently to the right and as I was watching it, the right engine bounced up about five inches, then back down almost a foot below its normal position, and then up again to a full foot over where it should be. In the light from the flames and sparks, I saw the top cowling being torn off. Then the whole engine fell away, out of sight while twisting to the right.

"Holy Toledo," or words to that effect, I called out to the skipper, "It's gone, it's [expletive] gone." He had identified the bad engine on the gauges, and not being able to see it from where he sat, he was trying to feather the propeller with no luck. All three control levers for the right engine were stuck and he could not move them, even with both hands. He went ahead and set maximum power on the left engine.

Having partly recovered from the initial shock, I realized it was about time to start flying this airplane again. I refocused my attention on the flight instruments. What I saw there gave me another surprise. In the three to five seconds that had passed since the initial "boom," the airplane had rolled into a 30-degree left bank. The ball in the slip-skid indicator was stuck in the left corner, and the airspeed had gone all to hell. From our cruise speed of 195 knots indicated, we were now down to 90 knots, well below blueline, and decelerating down toward VMC and stall speed at an alarming rate. I rolled the wings level and punched the nose down 10 degrees to regain control over the airspeed, which was the most immediate threat to our safety. I stepped on the left rudder pedal. This did not do us all that much good either, although the airspeed slowly crept up past 100 knots. The needle on the vertical speed indicator was pegged at the stop at minus 4,000 fpm, and the altimeter was unwinding to match. I was aiming for blueline in hope of regaining a positive climb rate, but never got that fast. The blueline (VYSE) on the Beech 99 is 127 knots. When we reached 120 knots the airplane was shaking, buffeting and vibrating so bad that I had difficulty reading the instruments. I feared that if I increased airspeed much more it would break the airplane apart. The VSI was still showing better than 3,500 fpm down. Brian had already set maximum power. There was nothing left.

Brian asked me what had happened. I said, "The engine is gone, it's not there, it's gone," and he said, "What? I've got to see this." He released his seat belt and got halfway out of his seat, leaning over me to see. I pointed my flashlight on the firewall, where the engine had been, to help him see. There was no fire now that the engine was gone and all we saw was the firewall, as if there had never been an engine there at all.

I had now stabilized the crippled airplane at 115 knots with a descent rate of 1,500 fpm, needing 70 percent aileron deflection to the right to maintain wings level and a good bit of left rudder to keep the ball near the center. Brian got back in his seat and called Anchorage Center, declaring an emergency and informing them that we were going to land straight in on Runway 21 at Fort Yukon.

I was not aware at this time that the instrument lights on both his VSI and his altimeter were burnt out, making them unreadable to him in the poor light. Therefore when he looked at his instruments all he saw was the horizon showing wings level. He was not aware that we had an alarming descent rate. Normally he would have no reason to worry; the 99 is a very good single-engine performer.

Based on this, his concern was whether there had been any damage to the landing gear and whether or not it would be a gear-up landing. Brian got out of his seat and told me that he was going to have a look through the back windows, and he went to the rear. We were now going through 3,000 feet.

Anchorage Center called, "F-BCH, what is the nature of your emergency? Do you have an engine failure?" I answered, "Negative, the right engine has departed the aircraft."

After what seemed to be a long time, Anchorage called again; this time the concern in his voice was very obvious. "BCH, how is the controllability of the aircraft?" I answered, "Not too good, we have max power on the good engine and we are going down 1,500 fpm."

At this time I realized that we were not going to make it to Fort Yukon. I made a shallow turn, left and south toward the Porcupine River, which was paralleling our original track and only three or four miles away.

Brian came back up front saying, "I can't see any damage back there." Then he looked ahead out of the windshield, realized that we were not at 6,000 feet any longer, and said, "Why aren't you maintaining your altitude?"

I answered, "I can't! We have max power on the good engine and gear and flaps are up and we are still going down 1,500 fpm. We're not going to make Fort Yukon. We will have to land on the river."

Brian got back in his seat in a hurry and said, "I'll take control"; I handed it over to him gladly. We were now at 1,500 feet.

I realized that we were not going to make the point on the river that I first thought. I said, "Turn toward the bend on your left. Stay away from those trees." I knew there was no way we would survive a crash in the trees with that descent rate, but we just might make it on the river.

One of my biggest worries was whether the vibrations and buffeting were going to break the airplane apart. "Please, baby, hang together just a little longer." We were now almost lined up with a spot on the bend of the river with only a few medium-size bushes.

Anchorage called, asking, "BCH say altitude passing through." I suppose we had gone off his screen. I checked my altimeter and said, "Five hundred, going down 1,500 feet a minute." Then I turned to Brian and said, "No gear, no flaps, right?" He only nodded his head: We both knew the last thing we needed now was more drag.

I realized that the landing lights were not on and reached over and turned them on, then looked out at what lay ahead. It looked rough and I was sure this one would hurt. All the money was on Brian now and all I could do was to hold on and hope.

Brian did not have an easy job. With all the aileron he had to use just to keep the wings level, he could not afford to lose any airspeed without losing roll control. Still our descent rate was 1,500 fpm, way too high. So all he could do was keep max power and fly it right down to the ground with the power at max. Two seconds before we hit, Brian chopped the power.

We hit hard on level ground and to our surprise bounced back in the air. Brian nursed the mortally wounded airplane back to the ground a second time (much softer this time). We slid on through a lot of driftwood and came to a stop on the river's edge.

We shut everything down in a hurry and got out in a dead run, only stopping to grab our parkas (it was −40° outside and we figured they'd be useful). We stopped some distance away, laughing with shaky knees, surprised at being alive, let alone unhurt. Neither of us had a single scratch.

Ten minutes after we hit, an aircraft arrived overhead. We ran back inside the 99 (there was no apparent danger of fire) and called it on the radio. The other pilot relayed our situation.

While we waited for a U.S. Army rescue helicopter to pick us up, we walked the length of our "landing path." At the first point of impact, we found the right-hand engine. It had been hanging by the control wires out of sight, and caused the incredible drag that pulled us down. And one of the propeller blades was broken off at the hub; the imbalance it had caused was what tore the engine off the mounts in the first place.

✈ ✈ ✈

*Given a choice, if the pilots had to "lose" an engine they'd have preferred it stay in place. If it was going to depart the airplane, it would have been much better if it hadn't tried to stay with them. Compared to a feathered engine, the hole left in a wing by a missing turboprop engine can be a real drag. Add a dangling engine and, unless luck steps in, you're on your way to a sure accident. The two pilots did everything they could and then had to trust their luck.*

*If the pilot in the next account didn't believe in his luck before his accident, he does now. Bad luck got him into trouble, but it was good luck that let him survive what should have been a final dead-stop landing.*

# Less Than Max Headroom
## by Terry Terrell

I was a line captain for a large and prosperous oil company in Texas that operated corporate jets as well as a fleet of twin-engine helicopters. Being cross-qualified, I found myself rewarded with a sparkling new Partenavia P68C Victor for my exclusive use within the company. Of course, the opportunity was not as glamorous as it sounds; it included the weekly patrolling of some 3,600 miles of inland natural-gas pipeline, running-mainly through the Texas desert wilderness.

The Partenavia was a joy to fly. My only criticism of the airplane was that the seat belts looked and felt as if they came out of a cheap car. As the weeks passed I settled into a pleasant routine of touring the Texas outback. This happy union between man and machine came to an abrupt halt, however, on a crisp VFR April morning. I was alone, enjoying a sea of bright yellow cactus blooms along our pipeline right-of-way. The next thing I saw was the hospital room where I woke up three days later.

I had flown, I was told, into the ground at a high cruise ground-speed from a very shallow angle of descent, and I hadn't flared or retarded the throttles at all. Then, apparently crumpled helplessly against the instrument panel, I'd decelerated through a thick stand of Texas mesquite. I had absolutely no recollection of the accident.

By the time I regained consciousness the FAA and the NTSB had made a thorough investigation of the wreckage. The airframe damage seemed categorically unsurvivable, and it was miraculous that I had come through with only a concussion and a broken nose. But I wanted to know what had happened. After all, I was a veteran of more than 6,000 unblemished hours, which included Navy jet and Coast Guard helicopter tours of duty, and my flying a presumably good airplane into level terrain, in clear weather, seemed impossible to believe.

During my convalescence my attending neurologist consistently maintained that I had been unconscious before the crash. Several uncomfortable months passed, seeing me medically recovered but professionally incapacitated by the troubling mystery surrounding the accident. Nothing in my extensive post-accident medical evaluation shed any light on the question of why an experienced pilot, with a previously excellent health history, should have lost consciousness and therefore control of a normally functioning aircraft. The NTSB and the FAA simply listed the cause as unknown.

In due course the company bought another Partenavia P68C, with mine, and assigned another pilot to a schedule that mimicked the one I had kept. Naturally, we watched his performance and listened to his observations and experiences very carefully. The detail we had been looking for came up quickly when, one day early into his tenure, he asked if I had noticed any problem with the seat belts in the P68.

"Well, I always thought they were like something out of a cheap car, and I rarely used the shoulder harness," I said. "In fact, I got into the habit of re-snugging the loose end of the lap belt many times during a day's flying because it constantly seemed to work itself loose."

"I'm having the same problem," he said. "It's particularly bad when the air is choppy down low. I keep bouncing my skull off the overhead."

Well, to say that that information was a revelation would be the understatement of the decade. I vividly remembered banging my own head into the cockpit roof many times in the rough air that was common to low-altitude flying over the desert, and then reflexively snugging the lap belt in protest afterward. It was not difficult to imagine that severe turbulence could land a disabling blow to the head of a poorly restrained pilot.

We discussed the "loose belt/rough air" problem within our department, and with the FAA and the NTSB, which ultimately asked us to write a report that was included in the accident file. It was the only plausible explanation for my bizarre mishap, and it gave me the necessary reassurance to resume my career.

✈ ✈ ✈

*Luck was at the controls of that Partenavia after the pilot checked out. It was Luck that kept the plane's wings level and flew it to a gentle—albeit high-speed—landing.*

*When the Husky pilot in the next story found himself trapped above a cloud layer, high over the mountains, with no power, he must have felt all alone. But apparently he wasn't. Luck was in his corner—and his cockpit.*

# Tall Mountains, Thick Clouds, No Power

### by Jim Wark

It was May, and I planned to fly my Aviat Husky from Afton, Wyoming, to Pueblo, Colorado. A frontal system with low ceilings was lying stationary along the route in the vicinity of Laramie. The sequence reports and pireps all indicated against VFR for the Laramie segment of the flight. My Husky, I thought, had a nose for finding the right path, so I took off, but after some time searching for a VFR passage none could be found. I headed for fuel and regrouping at Rawlins, Wyoming, where the FSS reported that there was little hope for a VFR flight unless I could go on top. Tops were reported at 10,000 to 12,000 feet.

There were breaks in the clouds to the north and west of Rawlins and I used those breaks to climb on top at around 12,000 feet. As the flight progressed toward Laramie, a continued shallow climb was needed to maintain VFR. About halfway between Rawlins and Laramie, when I was at 14,000, the climb started to seem sluggish and the manifold pressure seemed slightly low for the altitude. In the clear cold air carburetor ice was unlikely but with these symptoms, carb heat was in order. I applied full heat—no reaction; the manifold pressure remained constant. I puzzled over this for a few minutes and was hesitant to return the carb heat control to cold. Finally curiosity bested intuition and I pushed in the heat control.

It was as if the throttle had been closed. The engine went to idle. All engine instruments except manifold pressure were normal. I tried moving the throttle, mixture and carb heat controls—nothing had any effect. The engine was stuck at idle power.

Strangely, I didn't feel panic. Disbelief, yes; then a momentary feeling of the blood draining to my socks. At the time I was flying about 1,000 feet above the 13,000-foot cloud tops; I knew that 11,156-foot Elk Mountain was nearby, and that in most places the clouds went to the ground. For about a minute I tried to analyze the situation, convinced that there was some simple solution. Finally the "this is not a drill" message was received; I placed a call—"Rawlins Radio, Husky 2879T, I think I've got a Mayday" (leaving myself a little room to get out if things should get better). At the same time, still about 500 feet above the tops, I turned the Husky to a northerly heading that I hoped would take me away from Elk Mountain and headed for the lowest ground. I radioed the DME position to Rawlins and set the squawk to 7700.

Still without panic, but feeling a bit like a drowning man, I sank below the last sight of that lovely blue sky into darkening shades of gray. I have a lot of instrument time, but because the Husky and my previous two Pitts S-2s were not instrument airplanes, I had not flown even simulated IFR in 24 years. My Husky does have gyro instruments, however, and flying in the clouds was easy enough even for a novice pilot. Just hold a 350-degree heading and a 60-knot attitude, I told myself. With idle power this was giving me a 500-fpm rate of descent.

Continual cross-checking of position and chatter with Rawlins Radio and with another pilot who was offering some weather advice did cause my course to wander a bit and I was annoyed both with the airplane and with my skills. I carry Jepp RNAV charts so I had the Denver Center frequency at hand, but for some reason I didn't think of calling the controllers for radar vectors to the nearest airport until too late and too low.

The long idle-power descent through the clouds was a strange experience that will always haunt my memory. Physiologically, there was a calmness that was almost a numbness. My mind refused to even consider an unpleasant outcome for either myself or my Husky. I was straining to see the ground and don't remember ever looking at the altimeter.

Suddenly the ground was there! First a slight darkening, then faint shapes, then clearly from about 500 feet, Interstate 80 to one side (too far to reach); a dirt road, but not good (too close to power lines); another dirt road—very rough and muddy from two inches of rain, but that would be it. Wait! A paved road right under the wheels seemed to come from nowhere. It looked like the finish of an ILS approach. Touchdown was perfect, and the airplane didn't even get muddy. The road elevation was 7,300 feet.

Though the descent through 6,000 feet of cloud must have taken about 12 minutes, it seemed like a minute. The final 500-foot descent after I broke out VFR should have taken one minute, but it was as if time stood still; I clearly remember choosing two different landing sites before selecting the actual one. This completely improbable outcome, in an area where roads are few and big rocks and rough canyons are many, still leaves me wondering.

I pushed the Husky off the road and waited for someone to come by. In 10 minutes it was the sheriffs deputy who had been alerted to look for a downed airplane near Elk Mountain. We had to move the airplane and it was a long push to a suitable spot so I gave the engine a try. It started right up, but as the throttle was advanced it would first take power, then abruptly go to idle. As I managed to taxi to a safe place, a light slowly dawned.

After shutdown I removed the air filter, reached in the air duct and pulled out the loose carburetor heat air door. It is a flat plate that had broken from its shaft and was sucked up against the carburetor air inlet. A partially dislodged condition had probably been the cause of my initial concern at 14,000 feet. An AD note has since resolved the problem in the Husky fleet.

With a call from the FAA saying that "no enforcement action was warranted," I got a very good night's sleep. The next morning was clear and dry, so with no need for carb heat the Husky was fit to fly. The deputy made sure the road was clear and I was on my way.

*Why was there 500 feet between the clouds and the ground? Why in a mountainous area did a choice landing site suddenly appear directly beneath him? His training may have let him keep the rough side down, but no training taught him how to break out with maneuvering room and a landing site waiting for him.*

*In the next tale, a pilot again is forced to try to find a safe place to land in an inhospitable area. Instead of mountains, this pilot, in his A36 Bonanza, has to find a safe spot in the middle of California's suburban sprawl.*

# Deadstick Dilemma
### by Joseph Barber

We needed to be in San Francisco by seven o'clock that evening. I had calculated that the flight would take two hours, so we planned our departure for 3:30, giving us plenty of time for a pleasant VFR flight up the California coast. Five of us were on our way to a medical meeting. One of my passengers, Jean, had never flown in a small airplane before, but with my urging and assurances she had finally summoned her courage.

I took more care than usual on the preflight inspection because the Bonanza A36 had undergone its 100-hour inspection the day before. After one of those I always expect to find a loose fastener or a wrench where it doesn't belong. Finding nothing amiss, I carefully calculated weight and balance for the five of us, with baggage and enough fuel to fly to San Francisco with a safe margin; we were 50 pounds below gross at engine start. I seated Jean, Michael and Marjorie in the aft cabin, stowed the baggage and explained emergency procedures with the intent to inform but not alarm. I was glad later that I had had them practice opening the door from the inside.

I climbed through the cockpit door and moved over to the pilot's seat, and my friend Don climbed into the right seat. It was a hot July afternoon. Van Nuys Airport, amid the Los Angeles megalopolis, is one of the nation's busiest, so even on a Thursday afternoon we had several minutes to wait before finally being cleared for takeoff on 16R, the 8,000-foot runway.

The familiar sound of the Continental was reassuring as I opened the throttle and, even fully loaded on a hot day, we easily lifted off into the clear sky. As I began my right crosswind departure, I cheerfully called back to my three friends in the aft cabin, "Next stop, San Francisco." For their comfort I made my climbing turn smoothly and at a shallow rate as we flew above the city.

Less than a minute later, reaching 1,000 feet, I began to adjust the prop rpm for cruise climb and had the surprise of my flying life. The pitch-control knob had no effect on rpm. I didn't immediately know what this meant, but I knew I would have to return for landing.

As I began an immediate turn back toward the airport I called the tower: "Van Nuys Tower, 12 Alpha's having trouble with the rpm control. We're returning to the field."

"One Two Alpha, make a downwind to 16 Right and report abeam. What kind of trouble?"

"I've lost rpm control. We're not having any engine trouble at this time."

"Roger, 12 Alpha."

Now my instrument scan revealed a precipitous drop in oil pressure. Why had I not looked at the oil-pressure gauge when I first noticed a problem with the prop pitch control?

"One Two Alpha has lost oil pressure."

"Roger, 12 Alpha. Where are you at this time?"

"About two west, over the freeway."

"Are you going to be able to make the airport all right?"

"We'll see." Safe landing sites do not abound in the middle of the crowded San Fernando Valley. I continued to judge our chance of gliding to a landing back at the airport—or an alternative that I already had in mind, if necessary.

Was it better to keep high manifold pressure while I still had power to maintain altitude, or should I coddle the sick power-plant by reducing power, hoping to extend its life? With no relevant information to draw on, I decided to compromise and reduce power a bit. I thought I might make it to the runway because the engine was still running smoothly. Then I heard another pilot calling Van Nuys tower to relay the belated news that he had seen us trailing thick smoke after takeoff. I now realize we must have been losing oil ever since I had increased manifold pressure for takeoff power. As I listened, I wondered why he had kept this news to himself.

As the six cylinders each lost lubrication the engine began to vibrate and shake, to the point that I wondered if it might be torn from its mounts. Instead, it seized.

I had practiced engine-out approaches countless times over the years, but always from a safe altitude and with ample time to analyze the situation, choose a landing site and go through the emergency-landing checklist. This time, however, I didn't have much time to think. In the 40 seconds or so, that had elapsed since I had attempted to adjust the prop pitch, we had been heading toward the airport and had descended to 500 feet above the ground.

I had never before seen the blades of the prop stilled so abruptly. And it was so quiet. More unsettling was the unfamiliar and dramatic angle of descent needed to keep the airspeed at the 95-knot best glide speed. My view through the windshield was filled with rooftops and swimming pools. I had to resist the urge to pull the nose up to a more level attitude. In my mind I could hear the perpetually calm voice of my aerobatic instructor, Sammy Mason, saying, "You're doing fine, just keep the nose down. . . keep it down."

I knew we were close to making it to the runway, but now I did not think that we would. Rather than almost make the runway—and risk crashing into the airport fence—I decided we would land elsewhere. I could see that we had enough altitude to reach a farmer's field that had somehow survived the urban sprawl of the San Fernando Valley.

"One Two Alpha, would it be easier to make 34 Left?"

"One Two Alpha's not going to make the airport. We're going to land in the field in front."

"Roger, 12 Alpha. How many aboard?"

This question was alarming in its implications.

"There are five aboard."

From the right seat Don asked "We're going to be alright, aren't we?" "Yes," I said, mustering some reassurance.

It was almost a reflex for me to be aware of possible emergency landing sites wherever I flew, and because of its proximity to the airport I had often thought of this farmer's field as a possible site. Now I was actually going to use it. As we continued our descent, drawing alarmingly closer to houses and yards, I thought, "I'm not going to crash into someone's home." A Pan Am 727 had done just that the week before in New Orleans, and the awful scenes were fresh in my mind.

But we would be able to make the cornfield. Just like the barnstormers of days past, I would be landing in a farmer's field, deadstick.

Just as we were about to cross the last house-filled block, power lines loomed in front of us. I had had countless dreams in which I found myself in just such a situation, and now I had to decide; over or under? I decided to go over. I pulled gently back on the yoke, and at 85 knots we bobbled over the wires—I felt a tingling in my feet as we did so—and I put the gear down. A gear-up landing would really mess up the airplane, and the field looked smooth enough to handle the wheels.

I took this last moment to turn and tell my passengers to brace themselves.

Throughout the "approach" my attention was fully absorbed with maintaining control of the airplane and setting up for the landing in the field. It didn't occur to me to go through a landing checklist, and consequently I forgot a few things.

I had forgotten the flaps, so we sailed in ground effect just above the newly planted rows of corn for longer than I had expected. Finally touching down we rolled and rolled, and I realized I didn't know what hidden dangers (irrigation equipment, say) might lie ahead to trip us up. I gently nudged the brakes at about the same time the field became softer. We came to an abrupt stop and nosed over, the lower blade of the prop digging into the soft earth.

The moment we stopped, Michael had the aft cabin door open and was helping Marjorie and Jean out. As Don climbed out the cockpit door I called the tower to let them know we were safely down. As I started to get out I noticed I had forgotten to shut off the fuel and the master switch—two more items on that emergency-landing checklist that might have been vital if our landing had damaged the fuel lines.

The five of us gathered a few yards away, thankful for our safety and sand for the lovely 12 Alpha, now ignominiously resting nose-down in the mud. But I was elated.

After countless practice engine-out approaches (during which the hazardous part—the landing in a field—is never done), I had done the real thing. My hours of practice had paid off.

But questions arose in my mind. If I had not been so preoccupied with my passengers' comfort, might my instrument scan have revealed the loss of oil pressure earlier? Might we have turned back sooner, with enough altitude to carry us to a safe landing on the airport runway?

What I learned about flying is that engines really can quit, even when I am the pilot, even after a careful preflight inspection. A tiny crack in the oil-filter flange has allowed oil, under pressure, to bleed from the engine.

And I learned that altitude above us was not helpful. If I had climbed out at Vy that day, rather than in the more shallow climb I had chosen for the comfort of my passengers, I probably would have had sufficient altitude to make it all the way back to a safe landing on the runway.

Now I climb out efficiently on take-off, gaining altitude as quickly as possible. I no longer fly with the confidence that the engine won't quit. I do not enjoy flying low and slow, except over runways. And although I still enjoy sharing the flying experience with friends, I no longer attempt to convince anyone to fly with me.

✈ ✈ ✈

*Sure, if he had climbed quicker he might have made it back to the airport, but he didn't. The field was there, where and when he needed it. Let's hope it hasn't become an apartment complex the next time a pilot finds himself limping back to that airport.*

*Airplanes are designed to meet various strength limits depending on how they're to be flown. Aerobatic aircraft are designed to tolerate very high stresses. Rarely will an airplane break apart when it's flown as intended. Nevertheless, in the following incident, a pilot finds himself trying to nurse an airplane back to the ground after a structural failure.*

# Broken Wings
## by Guido Lepore

In our visions of shatterproof aerobatic pilots and airplanes, catastrophic failures just aren't expected to occur. Yet they do. I was practicing Unlimited class aerobatics in my modified Pitts Special directly over a large airport. Pulling out of a long vertical dive at over 200 mph, and hitting about eight Gs, I felt the airplane start shaking.

Initially, I thought something in the tail had let go, because I was having difficulty getting the nose back up to the horizon. I then chopped the throttle, thinking that perhaps I had lost part of the propeller and that the resulting vibration was going to rip out the engine. But when the nose finally came up through the horizon, there was no change in vibration, even with the throttle closed. Glancing out to the left, I saw the ailerons in rapid motion. Immediately I thought "aileron flutter," and kept pitching the nose up to about 60 degrees to lower the airspeed as quickly as possible. Scanning the panel, I saw that the airspeed indicator was reading zero, even though the left wings and ailerons were still oscillating wildly.

Not wanting to get into a tailslide situation, I leveled off the airplane and took an extra moment to assess the left wings. I felt sick to my stomach. The arrow between the wings' cross-bracing wires was missing; the upper aileron appeared to be in a full-down position, and the bottom one seemed to be flapping loosely. The wires were very loose and both the top and bottom wings were snapping up and down. In addition, the bottom wing seemed connected only at the rear spar fitting and was being held outboard with the "I" strut.

The wing had a rolling-type moment flexing forward and back as well as up and down. I suspected that the upper and lower ailerons were broken. Unknown to me at the time, the broken wing spar had pounded into the fuselage and damaged the pitot tube connection—the real reason I wasn't getting an airspeed indication.

I still didn't know how much control I had, but I knew that I had experienced a major structural failure. Now, I thought, is a good time to finally use the parachute that I had been sitting on all these years. As my left hand went up to the canopy latch, I looked over my left shoulder at the ground, entirely too close for comfort. The pull-up with power off had taken longer than usual, and I was now below 1,000 feet. Well, I thought, I've still got wings of sorts and the engines is still running. From the looks of the wing-flapping, I knew that it was only a matter of time before whatever was left holding them together was going to fatigue and fail. My dilemma: Should I use the unknown remaining wing-life time to climb up to a safer jumping altitude, or should I try getting the airplane back on the ground?

I decided to try to land. As I had been practicing flying an R22 helicopter in the infield during the morning hours, I knew that even if I didn't make it to a runway, that grass was going to be suitable. Coming back on the power, I descended, all the while experimenting with different attitudes trying to get the oscillation to stop. I was even more anxious in those last few moments because I was aware that if the wing fell off, I was too low to bail out.

Finally, I had to admit that the final meeting between the ground and the airplane might not be too pretty, even if I did manage some semblance of a controlled landing. Almost reluctantly, I pressed the mike button. "Mayday . . ." I began. The one word was all I got out. The airplane stalled. Without an airspeed indicator, I had gotten a little slow; now down to an altitude of about 400 feet the airplane started to roll left and pitch down. Full opposite aileron (the ones on the right side still worked) and up elevator had no effect. For a moment I thought to myself, "So, this is how it ends." Then the years of aerobatic experience offered another option. I added throttle, let the nose stay down and rolled the good wing down. Then, picking the nose up with rudder, I flew a descending turn to the left to stay over the infield and to get into the wind. I tried keeping a right slip to minimize the load on the left side, but the movement never stopped.

At the flare, wing movement ceased. In those final few feet, I ruddered the airplane into the wind and greased a landing on the infield grass. I hadn't even considered trying to swing around for a runway, even though there were three of them— one of which was 10,000 feet long. Rolling to a stop, I took a couple of deep breaths and then called the tower to request taxi clearance to the ramp.

On reflection, I realized that one of the improvements in the sport of aerobatics helped turn this near-catastrophe into a safe landing: locating aerobatic "boxes" in proximity to airports with control towers. This minimizes the fuel and time required to practice and also uses positive air traffic control to keep other airplanes clear of the area. The other major benefit, and the one I exercised this time, is having a landing surface directly beneath you and a facility with immediate emergency response capability. Even with the one-word "Mayday" call, the tower had immediately mobilized the fire trucks. Had I experienced a real problem impacting the ground, they would have been there within seconds. I'll never be comfortable flying aerobatics anywhere else.

*With the structural failure the Pitts Special suffered it should never have held together long enough to fly to the ground. If the pilot hadn't been practicing directly over the airport would the ending of this tale have been as satisfying?*

*The pilot who related the following account found himself the unwilling victim of one of the strongest forces of nature—one that could easily have had its way with his Robin.*

# Grabbed by a Mountain Wave
## by David Brown

The overall weather forecast looked good for a VFR flight in my Robin from Phoenix, Arizona, to my home near Long Beach, California. While being vectored out of the Phoenix TCA in bright sunshine, my only concern was a strengthening southerly wind that promised to provide turbulence from the surface to 16,000 feet. Once I was settled on course for Blythe, California, where I would stop to refuel, the turbulence made the ride uncomfortable enough for me to seek smoother air at 8,500 feet.

Although the wind was rising when I departed Blythe, the first hour following V-64 was uneventful. But as I neared the Palm Springs, California, area, the desert below was obscured by sand scraped up by the wind from the surface. When the Salton Sea appeared in the murk to the south, I contacted the Palm Springs controller. Weather was reported clear at my destination, Long Beach; but dust storms were restricting visibility around Palm Springs, and a cloud layer was hugging the mountains west of Palm Springs. Ahead, I could already see the edge of the cloud deck peeping over the mountains; Mount San Jacinto poked through the undercast to the right of my course, and another peak rose out of the clouds to my left.

While everything looked all right, I heeded an internal warning that there might be some rough air downwind of the mountain to the left of my course, and snugged down the straps of the aerobatic harness—feeling altogether ridiculous as the Robin hummed along under a brilliant blue sky.

Then *wham!*—it seemed like the little Robin had somehow fallen inside a washing machine. With a teeth-jarring jolt that knocked the wind out of me, the airplane slid sideways, banking, skidding and shuddering in violent turbulence. I was working hard just trying to keep the attitude somewhere near level. The tremendous forces first pushed me hard into the seat, then against the harness straps; once, going momentarily to negative G, my chartboard flew off my lap and out of reach. The Robin continued being thrashed through the sky for what seemed like an eternity. Then, just as quickly as it had begun, the turbulence stopped. The Robin flew on serenely. I took a deep breath, realizing that I had just passed through the rotor flow from a wave system.

Keenly aware of the presence of mountains under the cloud deck ahead, I told Palm Springs that I was climbing to 10,500 feet to avoid the turbulence. Palm Springs of course chose this time to tell me I was leaving their airspace and that I should "resume own navigation." So I reset the transponder to 1200 and checked the panel: engine instruments in the green. . . climb attitude and rate, good. . . airspeed 105 mph. . . altitude 8,300. . . altitude 8,100. . . altitude 8,000 and decreasing fast. The vertical speed indicator (VSI) now showed a 600-fpm descent rate.

My adrenalin was flowing. Here I was in the descending side of the flow from a full-blown mountain wave system—at full power. I pulled back on the stick to increase the climb angle, but this barely managed to arrest the sink. After a few seconds, the VSI showed a slow climb. Ten seconds later, things were at last normal. The Robin was climbing at 400 fpm.

The mountain was now behind my left shoulder and I started to relax as the Robin climbed through 9,000 feet in smooth air. Suddenly, the sink returned with a vengeance. The VSI now showed an 800 fpm descent. Although the Robin was still climbing at full power, the altimeter was unwinding and those clouds were getting perceptibly closer. A mountain ridge was hiding just under that cotton-wool deck of clouds. If the sink rate did not quit within the next few seconds, my only recourse would be to turn downwind to get out of the wave and dive back over the edge of the cloud deck into the valley where Palm Springs lay somewhere in a dust storm. Better, I thought, to fight the low-level turbulence than this invisible giant.

The air became as smooth as glass, when suddenly the altimeter began spinning crazily, and the Robin began climbing like a proverbial bat out of hell, passing through 11,000 feet, with the VSI needle pegged at a 2,000-fpm rate of climb.

I yanked the throttle back and pointed the nose down. The cloud deck was dropping away below at a rate more befitting that of a shuttle launch. Throttled back, in a steep dive, the airspeed was at the top of the yellow arc—and the Robin was still being propelled upwards like a leaf in a gale. Passing 12,000 feet, the VSI unpegged and began to show climb and descent indications within normal range. The OAT had dropped 20°, but I was perspiring heavily in my thin summer clothing.

After another mile or so, I ran out of the wave system. Luckily, I had been VFR. Descending and leveling off at 10,500, I headed for Long Beach and home—with a very healthy respect for mountain wave systems.

*The mountain wave posed a real and serious danger to the pilot of the Robin. In the next account, a pilot, flying freight in a Piper Lance, puts himself—and his airplane—at risk because of a danger he thinks is real. Luckily he didn't overstress his Lance during his maneuvering to avoid the King Air that wasn't.*

# Phantom Collision
### *by Robert I. Snow*

January 1976. The night is freezing cold but fantastically clear. Visibility is superb and lights can be seen 70 to 100 miles away. The mission tonight is to fly from my home base in central Connecticut to JFK Airport in New York. There I will pick up some priority freight for a regular client. It is 1:45 a.m.

I am flying VFR direct and am sincerely enjoying the flight. It is below zero outside at 3,000 feet but the heat in the Piper Lance works well and it is cozy warm in the cockpit. On this night the opportunity to fly more than compensates for having to be up at this hour.

I am still 40 or more miles out from JFK, flying along the north shore of Long Island. I have already monitored the ATIS and have been listening to approach control for the past several miles. I haven't called them yet; there is still plenty of time. JFK is landing on the parallel 4s and I can see a string of big jets facing me, landing lights on, either on final or being vectored. At this distance each jet's lights, emanating from two or more positions on the wings or fuselage, merge into one clearly visible white spot for each plane. I watch, mesmerized by the fact that I am able to see so far.

I don't remember if it was my pen, a chart or my sandwich that I was looking for when my attention was distracted from the scene ahead. In any case, I needed something from between the seats and I bent down for several seconds to find it. When I straightened up and looked out the windshield I had one of the most frightening shocks I have ever experienced in my flying career. There, right in front and coming at me, was a King Air or similar turboprop. His landing lights were on, one mounted on each wingtip, and I could tell by the relative motion between the two lights that he was banking frantically to his left to turn away from me. I had missed his nav lights as we came toward each other because my head had been buried in the cockpit. He had obviously just spotted me and had started evasive action, illuminating his landing lights as he did in an effort to wake me up to the danger. I grabbed the yoke and pushed, slamming against the seat belt. All manner of debris floated toward the ceiling as the plane plummeted. Later I would derive some compensatory satisfaction in the fact that I had dived. I had read somewhere that most pilots would instinctively pull if faced with the need for sudden evasive action. I had long ago vowed to push if it ever happened to me.

As I passed through 2,500 feet I felt a burning flush begin to suffuse my face. The King Air was not there; I had dived violently away from an optical illusion. Instinctively I looked around to see who might have seen me, as I sheepishly climbed back toward 3,000 feet. What I had seen as I raised my head were the landing lights of two jets on the localizer at JFK 50 miles away. Since I was approaching at an angle to the reciprocal of the bearing that they were flying to the airport, their lights were slightly staggered from my viewpoint with the nearer one appearing to the right of the one farther out. As I first looked out of the cockpit, the nearer aircraft had hit the glideslope and had started down. This created the relative motion between the two lights; two jets on the final approach, one level, the other descending.

What I'd thought I was seeing were the two wingtip-mounted landing lights of one aircraft, much closer than the two jets. That the light to the right was dropping had indicated to me that a roll to the left had been initiated. The illusion was so powerful that my imagination had filled in the space between the two lights with the shadowy outline of a King Air. I remember distinctly thinking that that was the type aircraft I saw.

Sounds silly, doesn't it? How can two aircraft 50 miles away be mistaken for one only several hundred yards away? If you haven't flown at night much, it is hard to understand, but believe me, there are many visual effects that can take you by surprise at night. These range from seeing Venus and believing it to be another airplane or a UFO (probably the most common night visual error in aviation) to something like what happened to me. If it ever happens to you, you'll be amazed at just how real the impressions can seem.

I am glad that there was no one with me who might have gotten hurt when I pitched downward, and that there was no damage to the Lance. I would have hated to have had to write on the incident report, "Cause of incident: evasive action to avoid collision with aircraft 50 miles away."

✈ ✈ ✈

*In the end it didn't matter whether the danger was real or only existed in the pilot's mind; it was potentially no less serious. Night freight pilots seem to have more than their share of gremlins riding along with them causing problems. The pilot of a Piper Navajo running freight in the San Francisco Bay area was forced to make an emergency landing when Lady Luck messed with his engine controls.*

# Oxygen On, Power Off
## *by John W. Dickson*

I'd been flying for a FAR Part 135 freight company for a little over a year and my Piper Navajo night freight run was a good time-builder. My logbook had just passed the 600 hours of PIC multiengine time and I was pretty confident. Of course, having passed three multiengine flight checks and just checking out in Beech 99 turboprops adds to the swollen ego.

The weather for the Oakland-Fresno-Burbank-Oakland trip called for IFR conditions in the Bay area later that evening, spreading to the Fresno area around midnight. What caught my attention was the forecast winds aloft, which predicted a great tailwind while en route to Burbank. I checked out a portable $O_2$ bottle, figuring I'd ask for 11,000 as an altitude for the FAT-to-BUR leg. (The company had opted for the additional payload and removed the built-in oxygen systems from most of their Navajo fleet.)

The preflight uncovered no problems and the cargo arrived on time. Soon I was airborne, climbing northeast from OAK on the standard IFR departure. The weather was still pretty good but I could see a low stratus layer covering parts of the bay. The sparkling lights of San Francisco, the Bay Bridge and Oakland made for a great sight. Savoring moments like this makes one appreciate being a pilot. ATC then turned me eastbound and soon I was established on V-244 and hoping for a "direct FAT" from the controller. However, the glimpse of an incredible number of air carriers stacked up between myself and my destination ruled that out. SFO obviously had some IFR weather on one of the busiest air-travel nights of the year. ATC kept me well clear of this swarm and the winds helped me to make the schedule into FAT.

I met the driver at the usual ramp location and was soon taxiing back out with a couple more bags and boxes. The clearance was ready and I had gotten my requested altitude of 11,000 feet. I was number one for departure at the runway end and I was soon airborne and climbing over the twinkling city lights of Fresno.

Established in the climb and on the airway, I had a moment to relax when I entered a smooth stratus layer. Soon, condensation and rain streaks appeared on the windshield. Pitot heats were on (it's a takeoff runway item) and the outside air temp still showed a reasonable margin above freezing; it would drop during the climb. The strobes were turned off, leaving the red glow of the left nav light visible in the mist.

Piper N6734 Lima, as a working airplane, wasn't the prettiest to look at. But the radios worked well and the engines ran smoothly with the gauges comfortably in the green. However, I cursed the autopilot pitch control; it still had a quirk and wouldn't hold an attitude during the climb.

Since it was night and passing through 5,000, I figured I'd get my $O_2$ mask plugged into the bottle and turned on. I had placed it on the floor behind the vacant copilot seat and against the wingspar. It lay facing sideways so that I could easily plug in my mask. But it wasn't that simple: it was pitch-dark behind the seat and I had to feel for the fitting to make the hookup. I couldn't quite find the right fitting and was trying to rotate the bottle around while hand-flying the plane. Of course, I had a flashlight, but I needed a third hand to hold it, fly the plane, and plug in the $O_2$ line. However, something was snagging the bottle; I twisted harder. The bottle finally rotated so that I could feel the fitting, got the hose inserted and I turned on the $O_2$ supply.

While I was putting on my mask, the right engine surged once, lost all power, resumed climb power, then did it again. The surging and yawing was awesome. I sat there dumbfounded and was totally confused. After using aviation's most popular phrase, I managed to switch on the fuel boost pumps, which didn't cure the problem. I checked the mag switches; they were on. The mixtures and props were forward, and fuel quantity was okay. But the problem continued. I thought some more obscene comments while lamely trying to figure it out. I checked the flight instruments: The ball in the turn coordinator was thrashing about in response to the errant engine and my fancy rudder footwork. Then I noticed the VSI; it was showing a steady 1,000 fpm down. My intended climb was not, and with my current altitude, this meant that I had about four minutes to figure out this mess, or the bank mail was going to be scattered over some farmer's field.

Something clicked in my befuddled brain and I called out the litany taught to every multiengine student—"Everything forward, everything up—identify—verify—and feather!" I finally stopped being a passenger and ensured again that the throttles were full forward, that the gear and flaps were up. I then retarded the right throttle and feathered the right engine. Then, in my best John Wayne future-airline captain voice (which sounded awful squeaky and tight), I advised approach that I had secured an engine and wanted the ILS back into Fresno. The controller quickly issued a vector for the localizer and a descent clearance. He then asked if I was declaring the "E" word ("Yes!"). Then he asked the obligatory questions about fuel ("lots"), passengers ("none"), and if I wanted the equipment. I figured that if I was working then I wanted them to be also, and said "yes" to that too.

I fished out the FAR RWY 29R ILS, which took valuable time, and quickly got it set up. I took an instant to check that the numbers were dialed in, and verified the localizer audio ident. The HSI needles slewed in the anticipated directions. "Good, we're above the glideslope and left of course," I thought. "Let's keep it above the glideslope." With a full cargo compartment, I was close to the max gross weight. Tonight, I didn't want to confirm Mr. Piper's single-engine-climb figures.

I made a few corrections and soon had the localizer and glideslope settled down. We broke out on about a three-mile final. The ceiling had dropped but it was a welcome sight to see Fresno's long main runway directly ahead. The landing was as uneventful as could be expected and soon I was trying to figure out how to taxi a loaded Navajo on one engine. I finally made it to the ramp, shut down the good engine and thanked the CFR crews. Then I called the company, which sent another airplane and a mechanic.

By now, it had started to rain and I was able to find a lineman to tow my broken airplane to the lee side of a hangar. Using a quarter, I was able to remove the engine cowling, then went in to curl up on a vacant FBO couch.

The other airplane finally arrived, the work was reloaded, and I described the symptoms to the mechanic. He grabbed his flashlight and climbed into the cockpit. Within a half minute, he said it was fixed. It seems that the right engine firewall fuel shutoff control snap cover was missing, and the oxygen bottle's carrying strap had snagged the exposed lever.

My one-armed wrestlings with the bottle had partially shut off that engine's fuel supply. When the engine lost power and rpm dropped, the fuel flow would drop enough for it to resume running. With the throttle set at climb power, the fuel available would be quickly exhausted and it would lose power again. Then the cycle would repeat itself. Boy, did I feel like a fool.

A couple of vital lessons were learned that night. With almost 500 hours of routine and reliable transportation in the Navajo, I had become quite complacent. If the problem had developed right after takeoff, I doubted that I could have properly handled it. I resolved to make every takeoff with heightened readiness and skepticism. The takeoff would also begin only after a mental review of the emergency numbers and single-engine procedures.

I also vividly learned that an engine can act up in a strange and disorientating manner. A real engine failure may be accompanied by unusual vibration, surging, yawing, noise, whatever. It can confuse you badly when you can least afford it. The smooth loss of power, administered with precision by company check airmen, may not adequately simulate what your left engine could be planning for you just after rotation. I was lucky that my experience happened with several thousand feet of insurance. You may not have this luxury—only your readiness and proficiency will determine the outcome.

# 3

# Their Master's Voice

Flight instructors are by definition the people from whom we learn to fly. They're the ones who introduce us to the skills and knowledge that give us the license to continue learning after we've graduated from their care and tutelage.

After an accident or incident, pilots often report having heard their instructor's voice coaxing and coaching them through the emergency, reminding them how to respond in the particular situation.

Sometimes they're correct. Sometimes, they're not. In the following accounts, pilots remembered—or misremembered—things their instructors had told them. As a result, some got out of trouble, but some got into trouble.

✈ ✈ ✈

## Bright Lights, Dark Night
### by Cindy Jesch

With 800 hours in my logbook, I departed on my first official cargo trip for my new employer. I was qualified to fly the Cessna 210 for day and night VFR, and this flight would take me to Las Vegas by day and back home to Reno by night.

The weather that day was perfect, and although it was forecast to remain VFR I filed an instrument flight plan for the return trip. I was feeling good as I flew into a moonless, coal-black night profuse with stars. After a peaceful two hours en route, city lights ahead to the west filled the windshield and I called the airport in sight. Because Reno had inbound ILS traffic for Runway 16, I was asked if I could accept Runway 25 for landing. No problem, I thought, I was already in position to make a straight-in to the west and had the runway lights in sight. I accepted the visual approach.

Somewhere in the depths of my memory, I recalled a lesson on night flying from my private pilot days. I seemed to remember an instructor telling me that, at night, if you had lights in sight, you were all right. In other words, I would clear any terrain ahead.

So I began a descent to the runway. It was a long final— about 10 nm. No problem. The gear was down, landing light on, four miles to go at 138 knots, when suddenly I was looking straight ahead at a Pinion pine. Maybe it was my imagination, but I swear I could have counted every rock and clump of sagebrush on that mountain.

With only seconds to react, I pulled up and turned left, in effect starting an instrument chandelle. By this time all I could see was black. I hollered a curse but quickly apologized to God because I was certain that I didn't have much time to be sorry before I hit the mountain.

As the airplane reached the top of the chandelle at about 70 knots, I realized that the only thing I had forgotten in the maneuver was to add power. With the nose about 15 degrees up into the blue half of the artificial horizon and the airspeed decreasing, I knew that I was also going to have to lower the nose as I added power. There are a lot of things I'd rather do than point the nose at a dark mountain that I can't even see. But I also firmly believe that airspeed makes airplanes fly; and my only chance for survival was to prevent the impending stall.

As I lowered the nose and added power, I looked over my right shoulder just in time to see the city lights shine again.

"88 Foxtrot, are you familiar with the terrain?" the controller asked as he lost sight of my landing light.

"Roger," was all I could muster. If I wasn't familiar before, I certainly was now. How could I have made such a stupid mistake?

My new boss wanted to know that too. He had been listening on the radio for my arrival. "How come the tower asked you if you were familiar with the terrain?" he asked. "You scared me. I went out looking to see how low you were."

"Scared you?" I replied.

He didn't pursue the matter further, and I was grateful for that. The lessons I learned that night have stayed with me and I have done a lot of thinking about the incident. Mountains are invisible on moonless nights. It is true that if the lights on the ground ahead are visible, you will clear the terrain—in level flight, that is. But in a glide, descent angle must also be considered.

Today, I have more than 5,000 hours in many different aircraft. I never fly into unfamiliar mountain airports at night without filing and flying an instrument flight plan. I do not accept visual approaches if an instrument approach is available.

If an airport doesn't have a published instrument approach procedure, I spiral down directly over the airport. I never make a straight-in approach to such a field at night.

The mistake that led to a near-tragic episode was one of misinterpreting a technique that I had learned during initial night training. Contributing to the incident was an overconfident attitude and the mistaken belief that I was familiar with the terrain. Fortunately, I am still alive to confess.

*"The mistake was misinterpreting a technique that I had learned during initial night training," the pilot said. She had remembered correctly, but applied the theory incorrectly.*

*In the next confession, the pilot again applies a rule stressed by his instructor. "The words of my instructor echoed through my mind," he relates, and then he dutifully applies the advice, only to scare hell out of himself and half the Royal Canadian Air Force training base. A word to the wise. . . .*

# Uncontrolled Finch
## by Hugh Gordon

I did my elementary RCAF flight training near St. Eugene, Ontario, at a small airfield about halfway between Ottawa and Montreal. We flew the Fleet Finch biplane. During my practice sessions on "circuits and bumps" I first grappled with the concept that what goes up always comes down and that if it has wings it should come down smoothly and gently.

"Now look," roared my instructor, "if you mess up a landing and bounce around too much, just open the throttle, go around and try again."

As practice proceeded I tried to keep that lesson in mind. Don't ride out a lousy landing. Shove pride to one side. Go around for another try.

My first solo was uneventful, the landing a neat three-pointer. Perhaps it would have been better had that initial circuit not been so smooth. After I had accumulated just one hour of solo time, orders came to practice more takeoffs and landings. The first circuit was flight-manual perfect. The Fleet feathered down nicely—so smoothly in fact that perhaps I felt just a bit too proud.

The second circuit also went well, and on the approach I had the aircraft aligned smack down the center of the runway. I was in complete command. The Fleet moved in obedience to my every touch. As the end of the runway slid under the yellow wings I began to flare for the final smooth three-pointer. Stick back. Hold it steady now. Steady. . . .

With all the surprise of an exploding bomb, my neat world disappeared. No longer was I flying a lovely responsive airplane; I was riding a wild, bucking monster that was trying its best to smash me to pieces. As I found out later, I had leveled off too high and the airplane had stalled, dropping its nose. The main wheels took the shock and began bouncing. The words of my instructor echoed through my mind: "If she starts bouncing, just hold the stick back hard and wait. If she doesn't settle down, open the throttle and go around."

I held the stick back hard. The Fleet bounced and bucked more violently. I rammed the throttle fully open. The engine growled to life and the violent motion stopped. I was back in control and a wave of relief surged through me.

When I raised my head enough to peer through the windshield my blood ran cold. My fingers circled the stick with a fearsome, white-knuckle grip. Directly ahead were a number of parked aircraft. My trainer had somehow veered 90 degrees from the runway. I could see a Fleet dead ahead and, on the other side of it, the familiar gasoline truck. Three men scattered like rabbits.

I swallowed hard and, during the few short seconds that followed, recalled something I had read about such situations. Stay on the ground as long as possible to build airspeed. If that isn't enough for flight, try to leapfrog over the obstacle and continue the takeoff run on the other side. I pressed the stick forward just a bit to delay takeoff until I was ready for my big leap over the fast-approaching barrier.

My eyes were glued to the gasoline truck and the parked aircraft ahead. When they were about 80 feet away and the airspeed read 70, I heaved back on the stick. The little airplane bounced off the ground.

I remember next staring down onto the parade square about 100 feet below. Men, practicing drill, stared up as I roared over. Climbing straight ahead, I waited until I regained some composure before attempting a turn. Then, carefully, I banked and circled back to the field. This time I silently prayed as the runway drew nearer, hoping there would be no repeat of the last show.

There wasn't. And as I climbed out of the aircraft one of the staff confronted me. "The commanding officer wants to see you," he said.

The CO sat behind his desk, my instructor standing beside him. Both wore expressions as if they'd just had their leaves canceled. The man behind the desk roared, "What the hell are you trying to do? You could have killed a lot of people with that maneuver." Carefully, I explained my actions, supporting them with what I thought was adequate justification. It was then that my instructor spoke up. "Do you realize, Gordon, that you were almost stopped when you decided to open the throttle?"

"No, sir."

The CO, now slowly calming down, rose from his chair and beckoned to me. "Step outside. There's something I want to show you."

Outside, he pointed. "There's the airplane you zoomed over, right where it's been for the past hour. What do you see in front of it?"

"My God," I said. "It's the control tower. How did I miss it?"

"That's what we wondered," my instructor said. "The men in the tower flattened themselves on the floor and prayed they might survive."

The CO continued the story. "Somehow, when you pulled back on the stick, you must have moved it to one side because you went to the left of the tower but below the top. You were level with the windows."

For a long moment I stared, then remarked, "But there's a hangar to the left of the tower and wires strung between the two buildings."

"Yes," replied the CO. "You went between the hangar and the tower. Your wheels just missed the wires. If you had touched the wires or either building you would have crashed into the parade square just beyond. Hundreds of men might have been injured or killed."

The three of us quietly stared at the space between the buildings where I had flown. My instructor broke the silence. "Gordon," he said softly. "Your guardian angel must have been with you. There's only 30 feet of space between those buildings." I paled. The wingspan of the Fleet Finch was 26 feet.

The CO continued, "If you bounce or have a rough landing, ride it out until it stops, for stop it will."

I have always remembered his sage advice. It saved my hide later in my service experience when landing a twin-engine aircraft. As it touched down, the airplane suddenly swerved due to a partially seized left brake. The aircraft swung off the runway and began bouncing wildly over the rough ground. I switched off the ignition and waited, knowing that the airplane would eventually stop—and stop it did. To go around may have courted disaster. Experience is a great but sometimes expensive teacher. But that's one way we learn about flying.

*The instructor's advice had been correct, but the pilot didn't realize the initial response had nearly solved the problem and the application of throttle was unnecessary. A rough landing very nearly became a tragic crash.*

*The advice of a "seasoned" instructor was brought home to the pilot of a Cessna Skylane when smoke in the cockpit got his attention during an IFR refresher flight.*

# Smoke Screen
### by Joseph Hogan

A wisp of smoke drifted by the artificial horizon. Then more. . . My God, the haze that I had been peering through was smoke . . . FIRE!

I turned to Lou, my instructor, who was helping me brush up on my IFR, and noticed her wide-eyed expression.

"We have an electrical fire," she said.

I quickly surveyed the cabin area, and through the spattered oil I saw we had just popped out of the cloud deck. *Spattered oil?* I didn't think to question it at the time.

I turned off the master switch, and Lou vented the cabin. With the radios' electronic displays dark, the chatter from ATC was silenced. We were now in what a friend of mine calls full stealth mode—no lights, no transponder and just a fading memory to Boston's radar. Only the hum of the secretly hemorrhaging Continental was audible.

Emergency procedures! We both had practiced these many times. Fly the airplane first, ready the fire extinguisher, get the airplane headed in the direction of intended landing, analyze and correct the problem (if possible), and let someone know what's going on.

As the cabin smoke cleared, we headed north with reduced power. Being familiar with the route of flight and monitoring our position, we knew Poughkeepsie Airport was about six miles to the north.

I put Lou on flame patrol with the fire extinguisher and searched for the hand-held transceiver. I rummaged around behind the seat for a while, then on the floor and finally found it. Next time I fly, that radio will be between the seats within easy reach, I told myself.

In the frenzied moments as we approached the airport, I jabbed wildly at the hand-held in order to set in the tower frequency. It is amazing what happens to the synapse processes under pressure. This radio is as easy to use as they come. I never thought that I could have so much difficulty setting a frequency. Finally I declared an emergency with Poughkeepsie Tower: no macho "I think we have a problem," just "Mayday Mayday Mayday! Skylane N9939E, five miles south, possible fire on board, smoke in the cabin, request permission to land."

We were promptly cleared to land on any runway. And then a barrage of questions: "How many souls on board? How many gallons of fuel on board? Will we need fire trucks?" A thousand questions of my own flashed through my head. Why was there oil covering the windscreen if the problem was an electrical fire? Why were we still trailing smoke like a dogfight casualty? If it *wasn't* an electrical fire, why did the smoke clear from the cabin as soon as the master switch was turned off? Then a chilling thought. . . if we were still trailing smoke, were there flames just out of sight licking the bottom of the fuselage?

Quickly I scanned the instrument panel for a clue. Oil pressure and temperature, normal. Cylinder-head temperature, normal. A quick scan under the panel provided no evidence of fire. A hand on the floor detected no undue heat.

By now I was in a real hurry to get on the ground. Too many unanswered questions. Lou called for full flaps as we ripped across the airport boundary fence high, hot and with a tailwind. I gingerly turned on the master switch, lowered the flaps and switched off the master again.

"Darn it, Lou, don't slip it so hard with full flaps."

She was not listening.

We oiled down every inch of an embarrassingly long runway that day. I fully expected a comment from the tower. But they only asked where we would like to park.

Once I finished kissing the ground, I discovered the source of the smoke. My freshly washed and waxed Skylane was soaked in slippery 15W15. A gallon and a half or better was dripping from every nook and cranny onto the FBO's ramp. The engine and exhaust stack heat were busy vaporizing oil. All in all, quite a sight.

After much huddling with friends and "experts," it was concluded that there had been no electrical fire. What caused all this trouble? An oil-cooler gasket worth $2.50 plus tax. It also was decided on good authority that the avionics fan must have drawn the smoke into the cabin and circulated it about. Luckily, our glide to the airport had not exhausted the oil supply, so there was no damage to the engine.

If the airport had not been within easy gliding distance, this little episode could have been *very* expensive; say, $10,000 for an oil-starved Continental straining to make it to an airport and, if unsuccessful, a little extra for a flatbed to haul the airframe out of a field.

Following the advice of a seasoned instructor, I never plan an IFR flight route that doesn't have at least 1,000-foot agl ceilings below. If something goes wrong, it's reassuring to have the visibility to choose the least hostile point of impact: preferably hard, smooth and no less than 1,500 feet long. To a reasonably low-time private pilot, familiar asphalt is worth a thousand farmers' fields.

Although some folks may say that insisting on 1,000-foot ceilings below is too conservative, the instructor who gave me that advice has been flying for some 30 years and has tens of thousands of flying hours without having to make one off-airport landing. I think I'll keep using his minimums.

By the way, that expensive handheld transceiver I bought myself paid for itself—in spades. Circling an inviting airfield looking for and trying to remember the difference between flashing green and steady green while the lifeblood of your engine is showering the neighborhood definitely wouldn't be any fun.

The more carefully I fly, the more I realize it's the details that zap you, if you let them.

Spend a few minutes and plan for the unexpected; the worse that can happen is you'll get an unwelcome bill and a story to tell.

*Altitude below is always better than altitude above, and when an emergency occurs in the soup, having comfortable maneuvering room below the clouds is obviously good insurance. Pilots have to arrive at their own personal minimums after balancing the risk involved with the importance of a particular flight.*

*In the following account, a student pilot learns both an important safety lesson and to appreciate an instructor who knows when he doesn't have to pound the lesson home. "How about a pair of Ray Bans?"*

# Asleep at the Stick
### *by Frederico Fiori*

The other day I was rereading *Fate is the Hunter* for the thousandth time. I was on the chapter where the author, Ernest Gann, then a copilot in a Douglas DC-3, pulled the landing gear up during the takeoff run before the captain's order. The airplane was barely flying. The tips of the propellers almost touched the ground, and the airplane hedgehopped until it built up enough speed to fly normally. What did the captain do? After landing, he turned quietly to his terrified copilot and said, "If you do this again, I will take you out of my will." Gann never forgot the lesson.

I was once a green copilot.

Having finished the helicopter commercial pilot course with the vast experience of 100 hours of flight, I was hired as a copilot. I had to accumulate 500 hours in a Bell 212 to be promoted to captain of a smaller helicopter, the Bell 206 JetRanger. With no training at all, I went from a modest two-seat trainer to an enormous 212 with 13 seats and two colossal turbines. I didn't even know how to turn on the radios.

Our job was to fly passengers and cargo to oil rigs in the Macaé Basin and Campos in Rio de Janeiro State, Brazil. One day we took off crammed with passengers, cargo and fuel for an all-day trip. It was summer and the sun was glaring on the horizon. After taking off, the captain moved his seat back and kindly allowed me to fly.

I was flying on a full stomach with the sun in my eyes. I was hypnotized and drowsy. Suddenly the captain jumped, moved his seat forward, took back command and told me quietly, "I have it. Take the checklist and call the control." I froze because I couldn't understand what was happening. I looked at the instrument panel and saw that the number-one engine indicators were slowly failing. The problem was simple. A fault in the fuel governor caused the number-one engine to roll back. When that engine's speed fell off, the other engine began to accelerate to keep rotor rpm at an acceptable level. But the good engine must be throttled back to keep it within limits. Naturally, alarms and warning lights go off before the worst happens. If you are attentive when scanning the panel, you can see the power loss immediately.

But that panel had too many instruments for my 100 hours. And I was inattentive. The captain lowered the collective to recover rotation, reduced the problematic engine, turned the governor to manual and accelerated the engine again, recovering the lost speed and altitude. Then he took a heading back to the coast. He tuned the radios and looked at me; I was still trying to find the checklist. During the flight back, I was desolate. What would the captain do? Would he throttle me with his own hands? Would he finish my career in the company? When we landed, he turned to me and said, "Your scan isn't the best with the sun on your face. How about a pair of Ray-Bans?"

Whips wouldn't have taught me so much. Seeing my embarrassment, he realized that I was conscious of how grave my mistake was. I bought a pair of sunglasses, but I learned much more than simply what a governor defect looks like, or that I had to scan the panel more efficiently. Just like Ernest Gann, I learned that next to experience, empathy makes the best teacher.

✈ ✈ ✈

*The message was subtle, but sufficient. In the following account, a freshly licensed pilot is reassured by his instructor that a proposed flight is not beyond his capability. "Piece of cake," he's told, but it doesn't turn out to be. Instead, he bites off a bit more than he can chew.*

# Buffalo Wings
## *by Thomas Criqui*

Shortly after I received my private pilot license through the USAF ROTC Flight Indoctrination Program at Ohio State University, two friends, Bob and Jim, asked if I would be interested in flying them to Montreal for a hockey game between the Montreal Canadiens and the Toronto Maple Leafs. This sounded like a fun trip, especially for a "hot stick" who had completed training in only five weeks. However, I suspected that a flight in November might be a little more than I could handle.

I worked up a tentative itinerary and took it to my FIP advisor, a B-52 pilot. His only comment was, "Looks good to me. Piece of cake. Sure, you're experienced enough to make the trip. Have a good time." That was all the convincing this hot stick needed, so I set the wheels in motion for the journey. Somewhere in the planning stage another friend, Tod, decided to go with us as far as Ottawa, Ontario, to visit a friend.

Before leaving for the airport that day I called the FBO, who assured me that the Cherokee 140 would be properly fueled for four people and baggage and ready for us upon our arrival. I then checked the weather with Columbus Flight Service and learned that there was a cold front slowly approaching from the west. Although the weather associated with the front was severe (heavy snow, low ceilings and visibilities), it was not due in the Columbus area or along our route until several hours after our scheduled departure.

When we arrived at the FBO we found that the Cherokee's tanks had been topped off, despite my request for half-full tanks, putting the airplane roughly 200 pounds above maximum takeoff weight with our load. The FBO could not find a way to siphon off the excess fuel, so I shot touch-and-goes free of charge for 45 minutes while the others waited.

Before departing I once again called night service to check out the progress of the cold front. What I heard was not encouraging. Although the front was still off to the west, it was moving faster than before and would be in the Columbus area within the hour. The ceiling had dropped in the past hour to 3,500 feet from 5,000 feet, and the winds had increased to 25 knots from 10 knots out of the west. I quickly loaded my passengers and we departed for Montreal at 5:15 p.m., one-and-a-quarter hours late.

We were initially vectored west for 10 miles before being cleared on course back to northeast. By 5:30 p.m. it was fairly dark and the ceiling was lowering ominously. Instead of the forecast 3,500 feet overcast, the ceiling was closer to 1,500 feet, and our groundspeed was 150 knots, for a whopping 50-knot tailwind component. The lowering ceiling, increasing darkness and strong winds were too much for this 45-hour private pilot, so I called Columbus Approach and requested the latest weather and radar vectors back to the airport. The news was not good, however. "Present Columbus weather: 300 feet overcast, visibility one mile in blowing snow . . ." I did not need to hear the rest.

My next thought was to divert to Cleveland before matters got too far out of hand, until I found out that it was closed: zero-zero in heavy, blowing snow. It seemed that the weather everywhere was much worse than forecast, so my next option was to continue on my original flight plan and stop in Buffalo to recheck the weather before continuing. Columbus Approach advised me that Buffalo was still VFR, with 3,000 feet overcast and three miles visibility in light rain, temperature 38° F.

After the first hour it looked as though everything was going to be all right. We were at 3,000 feet agl with about four miles visibility. Between the lights on the ground and the VOR stations, I thought I was aware of our position most of the time, although it was becoming more and more difficult to see the lights beneath us. I was really straining to see outside into the darkness, when suddenly there was a disturbingly loud sound of gushing air. I never before had heard this sound, so I looked around the cockpit, finally glancing at the instruments, which indicated that we were in a 30-degree banked, 15-degree diving turn to the right, with the airspeed indicator approaching the redline. Immediately I brought the aircraft back to straight-and-level flight, using the attitude indicator. Now I was flying only on instruments, which was something I had practiced for just 30 minutes the month before with my instructor in preparation for taking my private pilot check ride.

At FIP they taught us that in an emergency the first thing to do is maintain aircraft control. I had been a little slow, but now everything was under control. The second thing a pilot does is analyze the problem. We were at 1,000 feet agl, in cloud, at night, headed south. I brought the aircraft back to the proper heading and tried to figure out our position. The third thing that a pilot does in an emergency is take the appropriate corrective action. I wanted to land, but right now that was not a viable option, so I had to continue on to Buffalo and hope that the weather improved. My knees were starting to shake, and the sweat was starting to drip from my forehead as I realized that we were lost, in cloud at 2,000 feet msl, with the possibility of hills ahead of us at 3,000 feet msl.

I casually looked around the cockpit to see if my passengers had any idea of our predicament. Through the soft red lighting, I saw that my three friends were a picture of tranquility. Bob, seated to my right, was holding a magazine close to his face, so he could read it under the red glow of the overhead cabin light. Tod and Jim were both asleep in the back.

From the Buffalo ATIS I heard that the weather was 3,000 feet overcast with one mile visibility. Suddenly we broke out of the clouds and were directly above an airport. My initial sigh of relief was quickly replaced with the fear that I had managed to put us directly over Buffalo with all types of large aircraft trying to take off and land. I called Buffalo Approach Control for help and was relieved to hear the controller say that we were over Dunkirk, about 25 miles southwest of Buffalo.

Although the field was not VFR, and technically it was illegal for me to enter the Buffalo airport traffic area, I requested permission to land. The controller asked if we could make it to Rochester, New York. I replied that we could, but that if for some reason we could not land there, we did not have enough fuel to return to Buffalo. The controller said we could land at Buffalo and provided radar vectors to the field.

Once on final I had my first encounter with runway-centerline lights and touchdown-zone lights. At this point in my flying career I had no idea as to the function of all those lights that extended down an otherwise normal runway. Just to play it safe, I tried to touch down beyond all of them, but when I realized that the centerline lights extended the entire length, I set down about halfway down the runway. Thank goodness the touchdown-zone lights did not also extend the entire length of the runway.

The next morning we awoke to six inches of snow, still falling steadily. We went over to Buffalo Control Tower and visited for a while, and then we went down to the approach control facility, where I managed to meet and thank the controller who had brought us in safely the night before. I was in for one last surprise when he told me that the controllers on the previous night's shift had made bets as to whether or not we would make it in. Since no one thought we would, there was no bet. It had been only two weeks since their last encounter with another inexperienced pilot in the same predicament. He told me that the other pilot had bought the farm.

Every one of the controllers thought for sure that we were going to buy it, too.

Fortunately, we were luckier—and I do mean lucky, since there certainly was not much skill involved in taking off with the weather deteriorating. Neither was there much skill in passing up all those airports between Columbus and Buffalo. There was not much skill in failing to lean the engine at cruise, although in this case it would not have made much difference since the flight was at low altitude. There was not much skill in failing to turn around when we entered the clouds. No, there was not much skill in the conduct of this flight. But then how much skill can a pilot have with a grand total of only 45 hours of flying experience?

When we returned to Columbus, I went to see my FIP advisor. He seemed a little surprised when I told him that I had bitten off more than I could chew by trying to fly VFR from Columbus, Ohio, to Montreal in November, at night, and by counting on leaving on a particular day and returning two days later. After 8,000 hours of flying experience, I would not try that trip even now.

# 4

# Look Before
# You Leap

Airports were built so pilots and their planes could be assured of friendly landing sites. Unfortunately, we've become spoiled, and some of the techniques and skills that were common when off-airport landings were frequent have been forgotten or become rusty.

Because airports are generally well maintained, we get complacent when we're approaching them for a landing. We shouldn't be. Planes go where people put them, and if we put them into rough fields, wires, and each other, they're going to obey and go blindly ahead. We have to watch where we're going.

An NTSB accident investigator finds himself investigating an accident he was involved in, after he makes an off-airport landing in a Grumman Tiger. It was the same field where another Tiger had come to grief a week earlier.

## Short Stop
### *by Richard Shipman*

There I was—the official National Transportation Safety Board accident investigator—along with the safety representative from Gulfstream American, standing ankle-deep in freshly plowed Georgia soil and staring forlornly at the AA-5B Tiger we had just landed on this obviously unsuitable landing site. Only a week earlier a pilot had lost his life attempting a landing at this grass strip. Little did I realize that errors in judgment would jeopardize my safety during the subsequent accident investigation.

The fatal accident had occurred on a Sunday afternoon. The pilot was approaching the grass strip on his farm when, inexplicably, his AA-5B veered off to the right and crashed into a wooded area on the south side of the field. I was the NTSB accident investigator on duty, so I drove immediately to the crash site in rural Georgia. At the scene an FAA maintenance inspector and I documented the wreckage pattern, recorded available instrument and switch readings, examined the engine, took witness statements and conducted a normal on-site investigation.

But we could not positively account for all of the airplane's major aerodynamic components—necessary to eliminate the possibility of inflight breakup or structural failure. Because the airplane had cut a 300-foot swath through hardwood trees, the wreckage was badly broken up, and many of the metal pieces strewn along the wreckage path were difficult to identify. We decided to store the wreckage at a local airport and enlist the aid of the Gulfstream safety representative.

A week later I flew into Savannah, where the Gulfstream rep was waiting with an AA-5B to take us to the crash site. We planned to survey the area from the air and then land at the strip to reexamine the wreckage pattern. Then we'd fly to the local airport to inspect the stored wreckage.

I had already examined the landing strip in detail and found it completely suitable for a light airplane. The field was hard-packed sod, about 1,800 feet long, with no significant irregularities on the surface.

The flight to the crash site was uneventful. As we flew over the field, though, something looked different—I couldn't put my finger on what it was. I was flying at the time and suggested we make a low pass over the field before landing. We saw no major obstructions, but somehow the field still looked different from the way it had looked the week before.

We decided to go ahead and land. The Gulfstream rep asked if I wanted to make the landing, but since I had never flown an AA-5B before—let alone onto an unpaved strip—I turned the controls over to him. That was my only good judgment of the day.

As we turned final, a nagging doubt still plagued me. The surface of the field looked browner than I remembered, and not as smooth. Only when we began our flare could we make out the furrows in the soil, and by then it was too late. As soon as the mains touched down, we were thrown violently forward into our shoulder harnesses. The sudden deceleration pitched the Tiger forward onto its nosewheel, and for one terrifying moment I was sure we were going to flip over. The thought of the Tiger on its back in flames, with us trapped beneath the sliding canopy, filled my imagination. Fortunately the airplane came slamming back down on its main wheels and stopped in about 200 feet. As we jumped out of the cockpit—and sank up to our ankles in red Georgia soil—the reason for the short landing roll became obvious.

The field was freshly plowed, turning what had been hard-packed earth into soft, loose soil. We found out later that the widow of the deceased pilot, in a fit of grief, had directed her son to plow up the landing strip that had been involved in her husband's death. Fortunately, the only damage to the Tiger was some tearing of one of the wheel pants.

So there we were—two "experts" in aviation safety—staring at our stranded airplane and cursing our stupidity. By this time our dramatic arrival had attracted a number of local onlookers, including several farmers. I suppose situations arise in every pilot's career when it becomes necessary to ask some difficult questions: What airport have I landed at? Could I get radar vectors to my destination? Can I use your phone, Farmer Brown, to call the airport? We put it this way: "Gosh, we seem to be stuck in this field. Do you suppose there is any way you can use your farm equipment to turn this plowed field back into an airport?"

Before we knew it there was a string of tractors, trucks and cars driving up and down the field, packing down an area wide enough to take off from. We gave the takeoff careful thought, not wanting to compound our original mistake. I gallantly volunteered to stay on the ground to reduce aircraft weight as much as possible. We removed the wheel pants to prevent their catching in the soil, paced off the available takeoff area and consulted the handbook for performance data. When we were satisfied that a takeoff was safe, about 16 of us dragged the Tiger to the start of the "runway." After two abortive attempts, caused by one wheel or the other straying off the narrow strip of firm earth, the Gulfstream pilot was able to get the Tiger airborne, amid much cheering and ballcap-waving from the assembled crowd. He proceeded to the local airport with its 6,000 feet of paved runway, and I was driven there to meet him. We completed our investigation and also did a thorough examination of our Tiger. Then we flew back to Savannah, where I even got in some landings to check out in the Tiger. About a week later, the Gulfstream rep sent me an official-looking memorandum attesting to my qualification in the AA-5B, with a note appended stating that "soft-field landing techniques have been demonstrated."

Landing on unpaved surfaces is a way of life for general aviation pilots, but you have to know what kind of surface you're landing on. As we found out, fly-overs, even at low altitude, don't tell you nearly enough. And week-old information doesn't do you much good either.

The biggest lesson I learned was in humility. In my capacity as an accident investigator, I had routinely passed judgment on other pilots while subconsciously assuming that I was immune from such errors, given my substantial experience and my "official" position. But an airplane doesn't care about your résumé.

*The fly-over look at the soft field didn't show how soft it was. From the height of a low pass, it's not always possible to tell what the surface is like. Getting word about the latest conditions from someone on the ground should be very helpful. Providing, of course, he knows what he's talking about.*

# Ice Lander
## *by Richard Gamm*

It was one of those rare, bright, VFR midwinter days in Michigan. I had scheduled a cross-country business trip in my Cherokee 140 and planned to fly the first leg to a small grass strip to pick up a fellow safety engineer, then continue on to a commercial airport for our business appointment.

By my nature and profession I'm a conservative, safety-minded pilot. I grew up listening to my father's weather-related adventures flying B-17s around the Pacific in World War II, and I have never left anything to chance in my own flying. I knew I would have no problems at the commercial airport, but there had been a snowstorm the night before, and the 2,000-foot rolling grass runway might not be ready for my arrival. I decided to telephone the FBO ahead of time to see if the snow would be removed. The FBO manager told me the snow would be plowed long before my noon arrival.

I took off and headed for the small field located 60 miles southeast. Blue skies, calm west winds and unlimited visibility made the half-hour trip unusually pleasant. Arriving at the field,

I set up for a full-pattern approach. I called unicom to check on traffic, the active runway and snow removal. A teen-age voice came back over the radio, sounding tentative. "No airplanes, sir." The windsock confirmed my approach for Runway 27, but I called back again about the snow removal. "It's okay," the voice said.

As I turned final, I could see the runway and what appeared to be snowplow-cut edges. The noon sun was bleaching out the surrounding area. I set up for a full-flap, short-field landing. I made a normal flare, then at the moment of touchdown there was a deafening sound of ripping ice. Flying snow engulfed everything. The airplane came to an abrupt, nose-down stop.

Using my foot to push open the door, I stepped off the wing and into a foot of snow with half an inch of hard crust on top. The airplane's gear was somehow still attached and apparently in place. I called on the radio to the office, 1,800 feet down the runway, but received no answer. Surveying the situation, I decided it was possible to taxi to the office. After about 20 minutes I managed to snowshoe through and pulled into a well-plowed parking area in front of the FBO.

The shaken young man inside told me the boss had gone to lunch and that this was his first day operating the unicom. When the boss returned from lunch he explained that he had sold the airport to the county just the day before and that they had assured him they would clear the runway by seven the morning of my flight. It seems that I'd reached him at home when I'd called earlier about snow removal. He had assumed that the county would plow as promised.

My passenger and I drove his car to our appointment. When we returned to the field later that day, we found the runway plowed and a note from the road commission driver saying, "Sorry for your inconvenience." An airframe mechanic checked the airplane and judged it airworthy.

As I taxied down the hard-packed runway I decided that I would never again chance an unknown grass strip in winter— and I haven't since. Airplanes make lousy snowplows.

$$\rightarrow \rightarrow \rightarrow$$

*When everything comes to a stop, it's the pilot with whom the buck stops. The FBO thought the airport would be plowed, and the young man at the unicom was working his first day. The pilot relinquished his authority to those on the ground, people he felt were in a better position to know the runway's condition. How could he have known?*

*The runway isn't the only thing that can trip up pilots. The 172 pilot in the following account, who confesses after 16 years, had developed his own technique for landing—coming in fast and greasing it on. It worked well where there was enough runway, but when he found himself trying to plant it on a short strip, he forgot to look for the basic landing indicator.*

# Wind Socked
## *by John Emmerling*

I've sat on this story for 16 years. It's too embarrassing. But now, with the confidence I've gained as a 1,500-hour, reasonably competent Bonanza pilot, I'm finally willing to go public.

My wife and I had flown to Vermont's Warren-Sugarbush Airport on a sunny Friday afternoon in August. Then a 2,200-foot asphalt strip, the airport rose slightly to its midpoint and then fell off downhill. Several weeks earlier, with barely 100 hours of flying time logged, I had pioneered a new flying technique; smoother landings seemed to result if I just flew my rented Cessna 172 flat onto the runway. Why bother with that tricky flare, which sometimes left you hanging two feet above the runway when the airplane stopped flying? My slick, new, high-speed technique would surely rewrite the training manuals. And it worked just fine at my home base—Teterboro Airport, with its runways more than a mile long.

As I flat-greased it on that afternoon, the one-man FBO said to himself (he told me this later), "I probably should talk to that jerk about the way he's landing." But, being a true Vermonter, he kept quiet. We unloaded and met the couple we were staying with for the weekend.

My pal Alan—who was an advertising executive working on the Eastern Airlines account—had never been up in a single-engine airplane.

"Hey, yeah let's do it," Alan said enthusiastically to my suggestion that we go up for an early-morning hop. We finished our coffee, and at around seven a.m. we left our wives still snoozing in the sack and drove over to the quiet, dew-covered airfield. It was absolutely deserted.

I showed Alan how a professional preflight is accomplished, and we climbed into the 172. Unicom was, of course, unmanned, so I checked the windsock and decided to take off to the north. A brisk 10-knot breeze was coming straight down the runway.

Takeoff was routine and I was proud of the skill I was so obviously demonstrating. At about 500 feet agl we flew over toward the mountain and peered down at the condo we'd just left. There were no wives sitting on the sundeck with coffee. "Still in the sack, Alan," I said smugly. After a few minutes of sight-seeing, we headed back for the field. I called unicom for an advisory but got no answer. So I flew the pattern for the runway we had used 20 minutes ago.

My approach was high. I was over the numbers about 20 feet off the deck, but at least the airspeed indicator was reading my usual 70 knots. It did, however, seem that the ground was moving by a bit rapidly. Seconds went by. One-third of the runway was gone and the wheels were still not on the ground. I'd have no embarrassing go-arounds in the presence of my buddy, so I pushed forward on the yoke and forced the airplane down. We whizzed over the runway's midpoint hump and started downhill. I stomped hard on the brakes and could picture those incredibly tiny pads ("Don't forget to inspect these little brake pads," my instructor had always cautioned) struggling to stop my aluminum rocket. Suddenly I was very aware of the 30-foot pines off the end of the runway. A go-around was no longer possible.

Not to worry! The right brake faded and the airplane veered off into the grass on the right. We bounced along through some ruts, then hit a boggy section. The wheels mired. The airplane went up on its nose. The prop stopped quicker than I've ever seen one stop. And we were hanging from our seatbelts in an airplane balanced on its spinner. It was dead quiet for five seconds—then with a funny, wheezing, metallic sound the Cessna fell back on its tail.

"Is that how you usually land?" Alan asked.

The FBO operator arrived in time to witness the whole mess. The airplane had a bent prop, a blown nosewheel tire, a crinkled cowling and a squashed left wingtip. Alan was unblemished, and I owned the reddest face in the northeastern U.S.

During our brief flight the wind had shifted 180 degrees. After getting no answer on unicom, I had assumed that the active runway couldn't have changed. I didn't even glance at the windsock. And, of course, I had been high, fast and intent upon demonstrating my no-flare landing technique.

The Vermont FBO operator surveyed the damage—then looked me square in the eye. "You're supposed to look at windsocks," he said.

*Demonstrating his high-speed landing technique, the Cessna 172 pilot neglected to check the windsock and landed long. But it's not only general aviation pilots who are susceptible to high, fast landings. In the following confession, an airline pilot also finds himself hot and high trying to get a heavy Boeing 727 onto a runway he's neglected to brief himself about.*

# High Roller
### *by Eric Nolte*

Not long after I upgraded to the window seat of a Boeing 727 years ago, I scared myself into a clammy funk one dark and rainy night.

The night in question was at the end of the third day of a long trip, and every leg had ended with an instrument approach almost to minimums. A schizophrenic weather system had the whole East Coast in its grip. Everywhere there was unusually high pressure, 300-foot ceilings, visibility that averaged no more than three-quarters of a mile, and wind shear below 5,000 feet. Even in our heavy 727 we wallowed drunkenly through 30-knot airspeed fluctuations and severe turbulence.

The last two legs of the trip were to take us from Newark to Cleveland and back. By mid-evening, when we launched for Cleveland, the air was finally smooth, but we'd been on duty 10 hours and were as tired as a knock-knock joke. Still, the smooth air lulled us into thinking the hard part of our day was over. Twenty-five minutes from touchdown, we sat in our dark cocoon six miles high and began to prod ourselves into alertness for the approach. The flight engineer swiveled left and forward in his seat amid a clatter of metal latches and leaned forward between the pilots' seats to pass us the plastic card on which he'd just copied the Cleveland Hopkins weather.

"The Cleveland weather's crummier than forecast," he said. "Ceiling indefinite 500 feet, with visibility three-quarters of a mile in fog and light rain. Runway 5R/23L is closed, so they're using the ILS to Runway 28R. Winds are calm. Company operations tells me that a USAir DC-9 blew a tire on the long runway, and that's why they're using 28R." The captain and I muttered acknowledgment. It was my leg to fly, so I pulled out my Cleveland charts and, with the approach plate in hand, I briefed the crew on the procedure. I meticulously went over everything—the Jeppesen chart number, the date of the last revision so we knew we were all looking at the same plate, the navigation radio frequencies, inbound course, initial approach altitude, decision height and missed-approach procedure.

When the center controller cleared us to begin our descent from Flight Level 310, I disconnected the autopilot and flew by hand. Hand-flying from the top of the descent always makes me feel more in sync with the airplane when I cross the runway threshold, and even tonight I figured I wasn't too tired to do a good job.

But I *was* tired. I neglected to turn to another chart to find the actual length of Runway 28R, 6,015 feet of pavement that rises markedly for the first half and dips just as severely in the second. At the approach control hand-off I was kept high, consistent with the FAA's current slam-dunk philosophy of noise abatement, so as we descended out of 4,000 feet I was too fast and a full dot above the glideslope.

At the outer marker, five miles and a couple of minutes from touchdown, I was still too hot and high. I got the flaps and gear down a little late. By 500 feet agl I was not quite stable and still 15 knots too fast. I nudged the throttles just enough to spool up the engines because the old Pratt & Whitney JT8Ds take eight seconds to spool up from idle. We were carrying a full load of 185 guests in back—just below the airplane's maximum allowable landing weight.

I was still unaware how short the runway was. It wasn't really daunting, but 6,000 feet of runway in a 727 is comparable to 1,500 feet in a Cessna 172—nothing to raise your blood pressure, but you'd better be reasonably close to target airspeed as you cross the threshold at the right height, and you dare not float in the flare.

At 400 feet the clouds turned yellowish. Beneath the cloud deck at 300 feet I was still half a dot high on the glideslope, 10 knots above my target airspeed, and a little right of the centerline. Two melodies were playing through my head in counterpoint: the tenor was a string of obscenities directed at my uncharacteristically sloppy flying, and the bass a rational voice rehearsing the go-around procedure, if it should come to that— "Max power, flaps 15, positive rate, gear up. . ." I maneuvered more or less back into the slot and we crossed the threshold, still a smidgen high. The fog hid the shortness of the field.

The hump in the runway fooled me in the flare and we floated a few seconds while I fumbled for the pavement, which would be no cause for alarm in a DC-3, but at 140 knots you put roughly 200 feet of runway behind you for every second you spend hovering uselessly above the pavement. The glide-slope led me to the 1,000-foot marker on the runway. I was high. I floated at least four or five seconds despite chopping the power. Cutting the power normally causes a heavy 727 to simply stop flying. We arrived with the greasy side down on the right runway, and I yanked the engines to max reverse thrust and stomped on the brakes. At touchdown I noticed the hump in the runway, and then with the engines wailing at high power the white runway edge lights turned amber, indicating the last bit of the runway. A prickle of fear tugged at my throat. But I knew that the 727 is hugely capable of stopping quickly. I remembered a training film we saw in class that showed a bird at maximum landing weight coming to a full stop only a couple of thousand feet beyond the landing threshold. No sweat.

The captain, however, had a strange idea coursing through his brain. He was probably thinking, "Mustn't overheat the binders." The brakes on the 727, while enormously effective, have rather fragile brake-energy limits. One maximum effort, like a high-speed aborted takeoff, requires almost an hour's cooling period before you can take off again, and a few weeks ago we'd heard about an airplane that had actually set the brakes on fire after two aborted take-offs, even though the pilot had followed the book precisely.

As we slowed through 110 knots the captain said, "I got it." Of course, the captain always takes the airplane at some point on landing roll because he's got the only steering tiller for the 727's nosewheel. But now he had control of 75 tons of sizzling iron-mongery. I could see the runway's end rushing at us out of the fog faster than I'd ever seen it—and he was babying the brakes.

I couldn't stand it any longer and I began to "help" him on the brakes a little because we obviously weren't going to arrive at taxi speed before we went careering into the overrun.

"I said, *I got it*," the captain growled.

"Then *brake* this beast!" I yelled insubordinately. My eyes bulged wide, my heart was thumping away at the same power setting as the engines, which were roaring at near-maximum thrust when we reached the end of the runway. A new fog of dust was picked up by the reversers and swirled around our windows, and when the captain began to turn onto the last taxiway it was still too fast for comfort, but at least it was clear we'd cheated the Reaper. The foggy glow of the runway end lights flashed for an instant like fire in the dark crypt of the cockpit.

Later I remembered how, after the first few hundred hours of instructing, I thought I'd seen every mistake a student could make. Then I remembered the first student who really scared me with some new variation on an old theme, the student who drove home the wisdom that we *always* have to be vigilant around airplanes. I discovered that dark and rainy night that I have within myself the same virtuoso talent as the student who reminded me I had not seen it all; the same talent to cook up a new, and potentially lethal, concoction from the ordinary spices of our experience. *Bon appetit*, my friends, and may your blunders never cause you more than heartburn.

*The 727 crew learned the importance of knowing the runway length so they could plan their approach speeds accordingly. A Cessna 150 pilot, demonstrating emergency landing techniques, also neglected to become familiar with the "runway" where he was planning to land and ended up making a tricky "wheel" landing.*

# Trip Wire
### *by Victor Szilard*

The takeoff was uneventful. Then Terry asked me, "What happens if the engine quits?"

I am a 25-year-old commercial pilot with single- and multi-engine ratings and an airframe and powerplant certificate. I work as a freelance mechanic at the local airports. After a routine annual inspection on a Cessna 150, the airplane's caretaker gave me permission to fly it once or twice.

The 150 was kept on private property, and for years the owners had been using a stretch of unpaved county road as a runway. It was a half-mile long, with Highway 18 at one end and a telephone line at the other.

My new neighbor, Terry Knox, asked for a ride in an airplane, so I took him along. When Terry asked about engine failure, I explained that it is rare but that I'd be glad to simulate one and then demonstrate how to cope. We flew to an almost-dry riverbed that I had used as a practice area before. The only obstructions shown on my sectional chart were high-tension power lines that ran across the riverbed about a mile north of the Mojave River Fork Dam.

Reducing the engine power to idle, I set up for a normal glide into the riverbed. Just as I was adding power for the climb-out, there was a terrific crash. The airplane yawed violently to the left. I immediately kicked in right rudder, established a positive rate of climb to clear the dam and looked for a landing area.

We cleared the dam by 200 feet. I had no idea what we had hit or whether there was any damage. The wings looked fine. I turned to glance at the tail. The leading edge of the left horizontal stabilizer had a six-inch perpendicular gash. The tip was bent up and back, and it was touching the left elevator. A soft oscillation of the control wheel told me that it would hold long enough to get back to the strip.

Next, I looked out my window and down. The left main landing gear strut was gone. I asked Terry to look for the right main gear. "It's intact," he said. "Why do you ask?" I told him that the left strut was torn from the airplane—nothing to worry about. What else could I have said?

I envisioned a number of possible landing scenarios, none of which pleased me. I knew that if the left wing hit the ground at too high an airspeed we could flip or cartwheel. I tried not to think about the chain link fence that runs alongside half the length of the road landing strip.

I circled the strip for five minutes to plan my landing, hoping that somebody would see us and help if we couldn't get out of the airplane. On the downwind I told Terry to open his door, tighten his seat belt and hang on. When I cleared the power lines on short final I pulled the mixture to idle cutoff, closed the throttle, turned off the fuel valve and shut off the ignition, just as I had been taught. Then I slipped the airplane in on touchdown. The left wing dropped quickly. I added full right aileron. That gave my left wing a few seconds' more lift—time enough to steer the airplane onto the berm at the left side of the road. The airplane slid for a couple of hundred feet before the left wing caught the ground, and we groundlooped across the road. We halted against the opposite berm—15 feet from the chain link fence.

That afternoon we found the left strut in the Mojave riverbed, 200 feet from a one-inch, stainless-steel cable that runs across the river at about 50 feet. Both cable support structures were hidden by brush, and the cable itself was covered with a black coating. The U.S. Geological Survey sends a man out in a basket suspended from the cable to check the flow of the river. When I notified the FAA about the incident, they said they would try to put obstructions like that cable on future charts.

I learned some valuable lessons from this episode. First, don't fly at low altitudes unless it is absolutely necessary; I should have demonstrated the engine-out procedure at a much higher altitude, the way I was taught. And I learned that no matter how well you think you know an area, you should thoroughly check it before you fly into it.

✈ ✈ ✈

*It's obviously important to know what obstructions are in your path and to avoid them. The pilot of a Cessna 210 in the next account was watching where he was going, but the pilot of the Stearman behind him wasn't aware of the obstruction directly in his path.*

# Chomped by a Stearman
### *by J. Douglas Marshall*

My cousin and his wife were visiting from Montana and we planned to show them Guaymas in Mexico and spend some time on our boat there, fishing and lolling about. With our Cessna 210 all packed we were taxiing out from our shade hangar at Tucson International when we noticed a slow-moving Stearman entering the taxiway to fall in behind us. We were cleared to taxi to Runway 11R, the shorter of the parallel runways. Ground control ushered us and the Stearman across 11L. Approaching the yellow hold-short bars of 11R, our active runway, I turned left into the wind to do my run-up. I saw the Stearman plodding along and mentally envisioned him kicking his tail around to do his run-up beside us.

I was almost through my checklist when my cousin—himself an instrument-rated pilot—yelled at me, "He's not going to stop!" I glanced out my window. It was filled, at just about eye level, with a huge radial engine bearing down on us at a terrifying four mph. What followed was simply animal reaction spiced with a liberal dose of luck.

I had not quite turned left far enough on the taxiway to parallel the runway for my run-up. Because of the 210's angle to the taxiway centerline, the Stearman was aimed at a spot just aft of the left wing: a straight shot for that big prop to tear into the cabin. My hand was already on the throttle ready to run the engine rpm up for mag and prop checks, so when I saw the Stearman approaching I instinctively shoved the throttle to the firewall. Our Centurion moved about two feet before the Stearman hit—but that two feet, plus what little momentum we had achieved, undoubtedly saved all of our lives.

Those final seconds before impact were crucial in changing our story from one of disaster to a good hangar tale. My nosewheel was cocked to the left after my turn into the wind, so when our airplane moved forward it lurched to the left, becoming a more broadside target. First impact came when the Stearman's crossbraces between its left wings struck our left wingtip. This caused the Stearman to jerk to its left, instead of rumbling on directly into our passenger cabin. Instead of slicing up the cabin, the Stearman's big propeller started chewing on our left flap. I still had the throttle firewalled and our momentum, building with each second, carried both airplanes in a slow quarter-circle. Finally the 210's turbocharged engine reached max power with turboboost building, and we launched out of our deadly embrace with the big biplane. The surge of power carried my 210 into the dirt between the runways before I could yank the throttle back.

Everything happened so quickly and I was so engrossed in my evasive actions that I didn't have a chance to be scared. It was different for my three passengers, who had watched in horror as the Stearman's propeller came to within three feet of the passenger cabin.

When we got back to our FBO and inspected the airplane, we saw that the Stearman's prop had ripped to within inches of our left wing fuel tank. Not only had the cabin and its occupants come to within a few feet of being cut up like a pine log in a sawmill, but we had also been only inches away from having 45 gallons of 100LL be churned up and spit into two hot and running aircraft engines.

The moral of this story is to be sure to have God as your copilot—and always, but always, beware a taildragger, especially a big one, coming straight at you. Visibility over the nose of a Stearman and many other taildraggers is very poor while taxiing; you can't assume the taildragger pilot sees you. We were both on the same ground control frequency so it would have been very easy for me to have just called the other pilot to warn him that we were stopped just ahead of him. If you have doubts that another pilot sees you, in the air or on the ground, give him a call just to make sure. Next time, I will.

*Those moments waiting for the inevitable crunch must have seemed interminable. Time has a way of stretching out and slowing down when our senses are stimulated. The pilot of the Twin Comanche in the next account spent one of the longest moments of his life waiting to feel his plane settle to the runway. In that moment he moved from the column of pilots who will to the column of those who have.*

# Touch-and-Crunch
### *by Rick James*

Just about the time I expected to feel the tires touch pavement, I felt the airplane sink lower than I knew it should. A fraction of a second before the props hit the runway, I realized the gear was still up.

It had started out as a routine VFR flight from Waco, Texas, direct to Manhattan, Kansas. I was flying a rented Twin Comanche with my wife and two children on board. Everything along the way went smoothly, and the forecast for Manhattan was 25,000 feet scattered and 95° F—a beautiful day. Fifteen miles out, I called Manhattan Flight Service for an airport advisory. I was told Runway 21 was in use, with left traffic, and the wind was 200 degrees at 10 knots. As I approached the pattern, I ran my before-landing checklist. Descending to pattern altitude, we began to experience quite a bit of thermal turbulence. I moved the gear switch to the down position and mentally checked that the single green gear-down indicator was illuminated. I also completed the remainder of the prelanding checklist and turned onto final. When I began my flare, the airplane began to balloon upward, then settled to the runway.

The events of the next 60 seconds or so happened so quickly that I didn't have time to be scared until later. From the moment I realized something was wrong until the instant the props hit the runway must have been less than a second. As a fleeting thought, I considered applying full throttle and attempting to go around. I quickly discounted that because of two considerations. First, I was at VMC, if not below, as the props hit. Attempting to apply full throttle could have caused me to lose directional control of the aircraft. Second, because of the Twin Comanche's short props, I was only inches from the pavement when they hit. To apply power while that close to the ground would only have prolonged my slide, because I figured the fuselage was probably going to hit anyway.

Instead of applying power, I tried using rudder to keep the airplane in the center of the runway and lowered the nose, attempting to land as level as possible. The fuselage touched the runway and we slid just under 100 feet.

Later, a local A&P and I crawled into the cockpit and found that a gear-motor circuit breaker had popped. Evidently, this had happened some time after I first retracted the gear.

How did I complete a normal prelanding checklist and approach to the runway without noticing that the gear was not extended?

When I positioned the gear switch down, I mentally checked off the green gear-down light as illuminated. Now that I have been shown how the gear-indicator mechanism works, I am not so sure that it was activated at the time. The accident took place in the early afternoon on a bright, sunny day, so the sunlight through the window may have made the gear indicator appear to be lit when it actually was not. I really don't know. But I do recall checking off the indicator as illuminated.

The gravest omission on my part, however, was not checking the gear mirror on the left engine cowling, which provides a view of the nosewheel. Had I looked in that mirror, I would have clearly seen that the nosewheel was still retracted. This was pure sloppiness on my part.

Most pilots who fly retractables are used to "feeling" the gear come down. We usually notice the bump when it locks in place and the change in the attitude of the aircraft. Normally, I would have been suspicious when that was missing, but in the thermal turbulence the absence of those signs went unnoticed. Also, as I flared over the runway, I experienced excess float due to reduced drag, which should have made me suspicious.

As with most mishaps, this was a combination of several minor events that added up to an accident. The next thing I knew, I was rather sheepishly explaining to the owner how I managed to tear up his airplane.

Still, we were lucky. None of us got even a scratch—the only casualty was my pride. And in the future, you can bet that this is one pilot who will check that little mirror every time.

# 5

# With One Arm Tied Behind Them

Prudent pilots perform an adequate preflight inspection. They carefully go through the before-takeoff checklist and assure themselves that everything—at least at that moment—is working the way it's supposed to.

It's always been surprising how often good pilots, experienced pilots, pilots who know better, choose to go ahead with a flight even when there is an indication of a problem prior to takeoff.

In an emergency, we often tell ourselves, "This isn't really happening. . . ," but that's after we're in the air and need to get back down on the ground. Think how persuasive the arguments must be that convince us to go ahead with a flight when we know something's not quite right even before we release the brakes. Imagine how insidious and powerful the need or desire is for us to take the chances we do to get where we want to be.

In several of the incidents that follow, the pilots were convinced they had determined what the problem was and that it was not significant enough to cancel the impending flight or to delay continuing a flight after a landing. Out of little acorns grow great oak trees. . . out of small malfunctions grow serious failures.

## Turbo Meltdown
### by Truman O'Brien
The blackened spot on the runway was still visible when Jack showed up for his regular instrument lesson in his T-tail Piper

Lance. It had been almost a week since another Lance had crashed and burned. The pilot had reported smoke in the cockpit right after takeoff and had unsuccessfully attempted to return to the airport.

Working as Jack's instructor was a pleasant diversion from my duties as an airline pilot. It enabled me to keep my CFI current, and Jack was happy to have access to my experience. My experience, however, was about to teach us both a lesson.

Our preflight discussion that day centered around the earlier crash, particularly the proper procedure to follow when experiencing a problem just after takeoff. We departed Boeing Field, our home base, for Bremerton National, just 18 miles away, for some practice instrument approaches.

Jack's work was nearing check-ride proficiency, and after three approaches we headed back to Boeing Field. We were cruising above the Olympic Peninsula at 3,000 feet when I decided it was time to demonstrate the emergency procedures we had discussed earlier. We canceled our IFR clearance and, setting the imaginary field elevation at 800 feet, we started to climb. I then retarded the power to idle and Jack began a 180 back to the imaginary runway. We were both impressed when we nosed into the imaginary ground in a 45-degree bank. A valuable lesson had been learned. We picked up our IFR clearance and continued home.

We had just leveled at 3,000 feet when Jack said, "Hey, look at that." A thin wisp of smoke was rising out of the back of the instrument panel.

"Approach control, Lance 136 has smoke in the cockpit. We're shutting down," I said. I reached across and shut off the master switch, then took the controls. We were perfectly positioned for a high, right downwind leg to Port Orchard Airport, a 2,600-foot paved strip. On final we switched off all but one com radio, turned the master switch back on long enough to lower the gear, and got off another message to approach control. Then we landed.

I explained to Jack that this incident was almost identical with a radio fire I had had in a Cessna 210 a few years ago. Given the small quantity of smoke and the fact that it decreased instantly with our efforts, I was certain we had simply "smoked" a radio. A local mechanic inspected the radio stack; the DME was inoperative, felt hot and smelled of smoke. I was convinced that we had discovered the culprit. Following a complete run-up, we took off for home.

At 1,800 feet msl we were suddenly engulfed in dense, black, acrid smoke. This time I lowered the gear before turning off the master switch, and I surveyed the peninsula freeway below us. It was jammed with rush-hour traffic. The area all around us was densely wooded, which left only one option—the Port Orchard Airport, three miles behind us. Jack opened the vent window, covered his mouth with his handkerchief, and pulled out the fire extinguisher. We would be forced to land downwind and through the tall trees at the end of the runway. I pulled the mixture as soon as we touched down and threw the door open when we rolled to a stop off the runway. We crawled out of the airplane into fresh air.

Inspection of the aircraft revealed that the paint along the side and belly had been burned off, from the firewall to the wing leading edge. Portions of the aluminum skin and rivets had actually melted away.

When Jack's mechanic was flown in for a complete inspection, we learned that a clamp that holds an exhaust elbow of the turbocharger onto the muffler had not been safety-wired. The clamp had worked loose, allowing the 1,200° F exhaust to blast out of the turbocharger directly onto the firewall.

My experience with "smoked" radios had convinced me that our problem was of little consequence. Experience, then, is not always the best teacher, unless it is tempered with common sense.

✈ ✈ ✈

*Convinced that the problem that caused them to make an emergency landing was simply a "smoked radio," the pilots in the T-tail Lance took off for home. But the hot radio was only a red herring; the real culprit was still patiently waiting to ambush them.*

*The next confessor had noticed that one cylinder of his Piper Cherokee Six was running significantly hotter than the others. Although he consulted with other pilots and mechanics, the engine wasn't attended to before he set off on a flight early one morning. It was almost his last.*

# Bad Vibrations
## by George Noble
When I moved to central Idaho, I needed faster transportation. So I bought a 1965 Piper Cherokee Six and settled down to my new lifestyle.

The usual flight path from Boise to my home in Grangeville takes me due north up the Payette River Canyon to McCall, a town nestled in pine trees on the south shore of a large lake. I pass to the west of McCall, and continue north up the Little Salmon River Gorge to Riggins, where the Big Salmon River joins my path and we continue north. The river turns west about 20 miles south of Grangeville, so I begin my descent at that point. It's one hour, 15 minutes of beautiful scenery.

My problem began when I noticed that the EGT on cylinder number six was about 75° F higher than all the others. I asked four high-time pilots and two mechanics, one with years of experience, if they knew the implications of this temperature difference. They had no explanations, except that perhaps an injector was partially clogged, causing that cylinder to run very lean. I decided to seek out another mechanic.

At 6:15 one August morning, I departed Boise for Grangeville. At Horseshoe Bend, 20 nm from Boise, departure control released me, but I continued to monitor the channel just for the company. Soon the transmissions began to break up, so I switched to 122.8 to listen to McCall traffic, when the 300-hp Lycoming IO-540 began vibrating violently, and I experienced an abrupt loss of power.

My first thought was that I had lost a propeller tip, but then I thought about that injector nozzle. I enriched the mixture. No help. Then I performed all the standard emergency actions: checked both mags, turned on the fuel boost-pump, changed fuel tanks, and added full-rich mixture. When I increased the rpm with the prop control, the engine began to run smoother, but landing was the next order of business. I attempted to contact Boise Radio to declare an emergency, but I was out of range. So I turned the radio back to 122.8 and began to evaluate my options.

Just below me was a little airport at Cascade, and about 23 miles farther was McCall. I had been to McCall before and was impressed with the facility and the personnel; I would find a mechanic there.

Now, I'm a religious sort of guy. I believe that I didn't get to be 56 years old solely through my own efforts. Since the engine was running a little better, I said a short prayer in which I inquired whether I should push old 3214W another 25 miles to McCall. After all, my wife was faced with driving two hours as it was just to fetch me—if I could meet her in McCall it would save her 30 minutes on the road. The answer that I received was abrupt. It was as though someone took hold of my lapels and said, "Only an idiot would pass up a perfectly good opportunity to land right now." I chose Cascade; my wife could drive extra 30 minutes.

As I began my descent, I announced to Cascade traffic that I was experiencing an emergency and was about to make an approach to Runway 30. I elected not to shut down the engine; if I needed power for any reason, I would use it. Landing without further incident, I taxied to parking, shut down the engine and called my wife.

The following day a friend flew me down to visit my sick airplane. Our first task was to remove a spark plug. It had oil on it. We removed the lower plug, and it was damaged. No plugged injector here. A mechanic removed the cylinder and I got the bad news. A valve had cracked and broken, ruined the piston and redecorated the head. It took three weeks to get the cylinder rebuilt and installed.

I learned a good lesson from this experience. An anomaly such as an EGT indication considerably higher on one cylinder than on others demands immediate attention. I had flown 14W about an hour after noticing the EGT difference when the valve broke. A higher EGT could indicate a burned valve. The sensor in my airplane was reacting to the combustion temperature as combustion gases were escaping the cylinder. As the valve continued to disintegrate, a crack began to form, and failure was sure to follow.

I've a better appreciation for the airplane. Even though I'm an aerospace engineer, I didn't fully understand what the engine instruments were saying. I now listen intently to the engine and look at the EGT for each cylinder more frequently. I've found a better way of setting the mixture—and I monitor fuel flow and manifold pressure more closely.

✈ ✈ ✈

*Although he'd ignored the hot cylinder, the Cherokee Six pilot didn't ignore the closest airport and wisely elected to get the airplane on the ground as quickly as possible.*

*In the following incident, a cry for help from the engine, signaling that all was not well, was again misinterpreted and ascribed to a simple inconvenience rather than a critical problem.*

# Dangerous Diagnosis
## *by Jim Young*

We were in level cruise, about 1,500 feet agl, when our Debonair's newly rebuilt engine suddenly coughed and sputtered. I broke out in a cold sweat as I switched tanks, enriched the mixture and hit the fuel boost. Then the Continental IO-470 instantly smoothed out as if nothing had ever happened. I immediately started a climb to put a little more distance between Mother Earth and us, just in case.

We had had our wings clipped for nearly two months while our engine was rebuilt, and it had been finished for about a week when we took our flight. My partner had flown about six hours during that week with only a minor problem with the right fuel tank, which was immediately fixed. As I analyzed the situation, that fuel tank was most likely the culprit.

After a very high and careful landing, I checked everything from the tail forward; nary a screw was out of place. With everything so smooth after changing fuel tanks, I surmised the problem must have been that right tank. Famous last words.

After a couple of hours enjoying a little dinner, we headed back to the airplane for the one-hour flight home. We completed another lengthy preflight and were rolling at about 11:30 p.m. Approximately 500 feet agl the engine coughed and quit cold. Twenty-two years of flying and much careful practice didn't really prepare me for this. Trying to work the power controls, switch tanks and enrich the mixture while I started a 180-degree turn and got the nose down was about as much as I could handle. Whether we could make the airport was questionable, but it was sure preferable to the dark hills. About halfway through the turn I felt the airplane start to buffet and quickly shallowed out the turn and dropped the nose a little more. Just about then the airport lights came into view, Charlie dropped the landing gear for me, and I started the flaps down.

As we approached the threshold our Debonair was too high and too fast, but with 4,000 feet of runway available, the first thought of a happy ending to this began creeping into my mind. I should have known better. As we began settling over the runway, the engine suddenly coughed and sputtered back to life and pitched the nose up. While fighting the nose back down I realized that I must have inadvertently left the throttle in after my attempted restart had failed.

Power was the last thing we needed at this point, but grabbing the throttle provided another surprise I could have done without: it was stuck. As we ate up precious runway, I remember wondering how we could be so unlucky. We went from not knowing if we could make the runway to having too much power and no way to stop it. After fighting with the throttle a few seconds I organized my thoughts enough to kill the engine with the mixture, but there was very little runway left. We forced ourselves down on the last of the runway at about 90 knots, and only after considerable tire squeaking and wheelbarrowing did I manage to get the flaps retracted and use the brakes properly. We used all 4,000 feet of the runway and all of the overrun before we skidded to a stop within spitting distance of the airport boundary fence.

Subsequent inspection showed that the alternate-air door and attaching hardware had come off. The first instance of the engine coughing on the flight up was most likely the result of the small parts passing through the engine and out the exhaust, causing some damage to the cylinders and pistons. The engine failure was due to the loose alternate-air door being sucked into the intake and smothering the engine. Then, as the intake vacuum died with the engine, the alternate-air door simply fell out of the way. But with the throttle partly open and the blockage out of the way, the engine was ready to start running again. This time, however, the alternate-air door managed to jam deep enough into the intake to jam the throttle linkage as well as kill the engine.

You can bet your bottom dollar I will never guess again what might be wrong with an engine. That's what mechanics are for. When lives are at stake, we should leave no room for error.

✈ ✈ ✈

*The engine coughed when it swallowed small pieces of the alternate-air door, but it wasn't until after the door came loose and blocked the air intake and smothered the engine that the pilot realized he had ignored the warning signs.*

*In the next confession, an Air Force pilot, flying an OV-10 Bronco from Thailand, chooses to disregard the engines' efforts to warn him that they don't want to go to war.*

# Not Up to Power
## by Bob Berry

Even for Thailand it was hot. Roger and I had started sweating the moment we stepped out of the air-conditioned operations building of the "Rustic" FACs (forward air controllers). We weren't helped at all by having to wear boots, long-sleeved Nomex flight suits, survival vests with about 30 pounds of gear apiece, and two canvas sacks full of maps.

"It's a good thing airplanes don't sweat," quipped Roger, "otherwise them Broncos woulda melted by now." Roger was a good man, hard-working and aggressive. He was referring to our mount, the Rockwell OV-10A. The OV-10 was my first assignment out of USAF pilot training and I felt very comfortable with the airplane and the mission. I had been flying combat for six months, Roger for nine. Roger had been driving a fuel truck in Bien Hoa before being assigned to the Rustics. His main job was interpretation: Our "beat" was Cambodia, and very few Cambodians spoke English at that time, so we carried "GIBs" (guys in back) to speak French with them.

"Well," I groaned, "if they'd shrink to Shetland pony-size I'd be happy." The Bronco was a big ship, and mounting the boarding step burdened with gear was a job for the young. Standing on the step stowing my gear, I watched Roger tumble into the rear cockpit. He was 18 years old and a seasoned warrior. I was 24. Roger thought me "old." We worked together very well.

As I settled into my office, the crew chief mounted the step to help me strap in. "Don't bring back no rockets or ammo today, lieutenant. Hit 'em hard for me."

"You bet," I said, "Thanks for the help, chief." The Bronco responded to my manipulations and soon I had two Garretts turning with all needles in the green. Taxi clearance to the arming area was obtained, and five minutes later we were parked with both hands on our respective glareshields as the armorers removed the pins that turned our guns and rockets into deadly weapons. With arming complete, we were cleared to Runway 25 at Ubon RTAFB.

The Bronco had two 715-hp Garrett engines, but we routinely operated at weights exceeding 13,000 pounds at takeoff. This made consulting the takeoff power charts an absolute necessity. At those weights the airplane was only marginally powered at best. The charts were entered with OAT and PA and produced a minimum acceptable torque setting, measured in lbs/ft. At maximum power any setting less than the chart value called for an abort. I checked my chart, and wrote the magic number on the canopy with a grease pencil.

A short exchange with Roger confirmed we were both ready to go, and when I called, Ubon Tower cleared us for immediate takeoff. I turned onto the centerline, rolled forward a few feet to straighten out, stood on the brakes, and advanced the throttles. When the engines came to full power I checked the torque meters. They were reading significantly below the chart value. I argued with myself: Abort, or go ahead? One of my squadron mates was waiting for me to relieve him, there's a war going on, and besides, the runway was plenty long. I convinced myself it was safe to go on.

I released the brakes, bent the throttles over the stops, and hoped for salvation. Acceleration was decidedly under-whelming, but I was soon reassured by a satisfying "thump, thump" as we rolled over the approach-end arresting cable. There was still considerable concrete in front of us and the airspeed indicator had just come off the peg. Too soon that comforting stretch of concrete was behind us and we weren't anywhere near flying speed. Panic took its shot but quickly retreated. I was too concerned with the wire fence that Roger and I would be charging through unless something happened soon.

The "thump, thump" of the departure-end arresting cable shook me out of my concern with the fence, and for a reason I still don't understand today I reached for the flap lever and dropped the first notch of flaps. I heaved back on the stick and wished mightily for an immediate confluence of the forces necessary to make this airplane fly. It occurred about two seconds later. The nosewheel bounced twice and then remained airborne, climbing slowly to the proper takeoff position. The main gear struts clunked as they extended to their full length and daylight appeared between the tires and the runway. I snapped the gear handle up and leveled at 10 feet agl.

I knew we would clear the fence now, but a new problem was dead center in the middle of the windscreen: Sofia's, a local "business establishment" and one of the tallest buildings in Ubon, sat directly off the end of Runway 25. It was no problem normally, but now wasn't normal. Life depended on airspeed now. If we gained enough to make a turn possible, we would live. Otherwise, we would become part of the local folklore. Moving the stick to the right resulted in a momentary shudder and a slight right bank. It was enough however, and as Sofia's slid off to the left, Roger and I could see the faces of Sofia's employees pressed to the windows. I quickly leveled the wings and let out the breath I had been holding for too long. Two minutes later the airspeed was close to normal and I signed off with Ubon Tower. Roger thanked me for the low-level tour of downtown Ubon, and then he and I went off to war.

Five hours later we were back. Except for the takeoff it had been a good day, and we brought home no rockets or ammo. I did, however, learn a lesson that has stayed with me since that day in 1971. Not using a checklist is foolish, but using one and then failing to heed its built-in warnings is doubly so. I decided I would have plenty of opportunities to look, sound and act foolish outside the cockpit.

Flying can be a fun game, but certain phases of flight should always be treated as the deadly serious business that they are. Not every takeoff can be made, and when the numbers say go back inside, I advise you do so. I used up approximately 10 years of good luck on that one takeoff. On every takeoff since then I've replaced luck with preparation and a belief in the truth of the numbers.

*"At maximum power any setting less than the chart value called for an abort." That's certainly straightforward. But he didn't abort. The mission seemed too important. How important would it have seemed if they had plowed into Sofia's?*

*"What ifs" often help pilots realize how close a close call really was. In the following account, two pilots go up to do spin training in an airplane that really isn't approved for it. Later they wonder "what if" they had needed to get out quickly.*

# No Way Out
### by Henry Hutchins

"I've got it!" Don yelled as he took the airplane. He neutralized the ailerons, applied opposite rudder and held his breath. The airplane continued to spin. When we passed through 2,500 feet, I slapped the D-ring on my parachute just to be sure I remembered where it was, and reached for the overhead latch. . .

Don had not wanted to fly Three-Seven Charlie, let alone spin it. I had insisted that I needed spin training and that Don, as my instructor, should come along. He suggested that we venture across the field and rent another airplane to do spins.

"But Don," I persisted, "the company has provided an airplane. They want us to do service tests on Three-Seven Charlie."

"Three-Seven Charlie is a *prototype*." Don's voice was tinged with exasperation.

I knew that he was close to giving in, so I just leaned against his office door and waited.

"Okay," deep breath, "you go over to flight test and borrow two parachutes. Let me know when you're ready."

"All right!"

Thirty minutes later I was back with the parachutes. "Let's go."

"Not so fast, hotshot. Have you ever made a jump?"

"Well, no. But. . ."

"But nothing." Don had lapsed into his instructor voice. "I have made two jumps. The first was a static-line jump with an instructor so I would know what to do if I ever had to use a parachute. The second time I was flying a new homebuilt with the owner, who tried an inverted maneuver that broke the main spar."

"So, no problem. Right?" I asked hopefully.

"Wrong. I almost got stuck climbing out of the small cockpit. Just practicing with the parachute was not enough. You and I are going to take the next half hour learning how to get out of that airplane."

"Great!" I said, picking up my parachute and following Don out the door and onto the ramp.

Three-Seven Charlie waited on the tarmac—a white Piper Tomahawk with a narrow strip of red painted the entire length of the fuselage. It looked like the other airplanes that would follow it, but there was a unique difference: Three-Seven Charlie was the service test prototype—not the flight-test model that had been flown hundreds of hours for certification, but the one that was pulled from the airplane's first production run to determine that the tooling was correct. As mistakes were found in this airplane, the entire production run would then be corrected.

We strapped on the parachutes and climbed into the cockpit. The seats had to be pushed all the way back to allow for the burden on our backs. We made sure we could both reach everything in the cockpit and then opened the doors and rolled out onto the wing. Once on the wing, I rolled off the trailing edge and landed on my feet.

"One thousand, two thousand, three thousand," I yelled. "Geronimo!" I slapped the D-ring, pulled my hand away and looked at Don across the fuselage. He was not smiling.

We did it again. This time I acted more serious. What I felt was pure excitement.

"Okay," Don said at last, "I guess we're as ready as we're ever going to be. Let's go."

We took off and began climbing. Don wanted 7,000 feet below us before we spun the airplane. It took a long time to climb that high, and we passed the time with small talk. As we passed through 6,500 feet, Don turned serious. He went over the procedure we would follow to stall the airplane and the actions required to recover.

"I'll do it first. You follow me through."

I nodded my head and held the yoke lightly, feeling the gentle pressure he used to ease the nose up just above the horizon.

Suddenly we were upside down! The nose dropped through and the ground rotated beneath our spinner. The earth slowed, then stopped, and Don eased back on the yoke to bring the airplane level.

"Are you all right?" he asked.

"Yeah, sure," I replied, not feeling that way at all. "What happened?"

'This time you take it." Don leaned away from the controls. I scanned the panel, pulled back on the throttle and held the nose up. Once again the Tomahawk rolled inverted, then fell through to a spin. A steep spin. After only a moment of disorientation I slid the yoke forward and applied rudder. The airplane obeyed nicely.

"You lost 2,000 feet," Don said sternly. "I'm glad you didn't do that in the traffic pattern. Let's climb back up and try it again."

As we climbed I compared this airplane to others I had spun. They had all just dropped their nose and turned around and around. This one tossed me upside down first, then spun nearly vertical. This time I would be ready.

"Let's spin to the right this time. Hold that rudder." Don was looking for traffic as he spoke.

I held the rudder, saying to myself, "This time I'm ready."

I was not ready. The snap roll to the right threw me off the seat. I scrambled to grab the yoke as my feet searched for the pedals. I neutralized the ailerons and pushed opposite rudder. Nothing happened. I pushed the controls the other way and put them back where they should be. Still nothing happened. That's when Don yelled, "I've got it!"

I watched him do the same thing I had done with the same

result. I checked the altimeter as it swept past 2,500 feet. When Don reached for the flap handle, I knew it was time to leave. As I reached for the overhead latch, I floated from the seat and bumped my head on the headliner. From there I crashed back into the seat as the horizon slid into place just over the nose.

"I've had enough for today," Don said. "Let's go home."

"Okay with me," I said weakly.

A few minutes later we were taxiing to the ramp and calmly discussing the airplane's spin characteristics. I shut down the engine and reached for the overhead latch. It would not move. My body went cold as sweat beaded across my forehead. "Don, I can't open this."

Don reached up and tugged at the latch. His body stiffened, he started to speak but did not. We both realized that we had been trapped inside this airplane with no way to get out. Our parachutes and preparations had been useless.

We used the radio to get someone out to the airplane who could pass hand tools through the small window. We removed the headliner and disassembled the overhead latch. As soon as we got out, I found a phone to report what had happened to us. That was when I discovered what problems Three-Seven Charlie had helped to solve.

The wing had a slight warp that caused it to stall suddenly. This had been corrected with a stall strip on all the production airplanes. The first design of the overhead latch was faulty, but the new design had been installed in the production fleet, except for Three-Seven Charlie.

Both Don and I flew Three-Seven Charlie again, when no other airplane was available. Neither of us found it necessary to stall the trainer again, even after the stall strips were added. The headliner now had a small modification, made by a pilot with a pocket knife, that allowed several fingers to be inserted to trip the overhead latch.

Most of what I've learned tends to fade with time and disuse, but the lesson Three-Seven Charlie taught me is as clear now as that terrifying moment I learned it nine years ago: after you plan for everything that could possibly go wrong, something else will.

✈ ✈ ✈

*The two experienced pilots were very confident in their ability. They knew they were flying a prototype that didn't meet produc-*

*tion standards but they weren't worried. Even if things went ter-
ribly wrong, they thought they had an out—through the canopy.
What a shock to find out that their chutes only gave them a false
sense of security.*

*The rush to get a Cessna 206 floatplane through its 100-hour
inspection and up to acceptable limits should have given the
next pilot cause to worry. But bush pilots don't sweat the small
stuff. It took a growing puddle of fuel on the cockpit floor, a
strange lack of power, and sleet and turbulence to get the 206 pilot's attention.*

# High Explosives
### by William Haines

The air taxi's dispatcher had scheduled me to fly as soon as the
Cessna 206 floatplane came out of its 100-hour inspection. A
compression check, however, found four of the six cylinders
below limits. With most of the company's aircraft still in winter
storage, the mechanics worked the rest of the day, well into the
night and all the next morning to get 75Q back on the line.

It was after one p.m. when I finally took off from the company's Juneau, Alaska, base, bound for two small communities
and a coastal cabin. Wet snow clung to thickly forested slopes
that rose and disappeared into dreary stratus cloud—a typical
scene for coastal Alaska in early April.

After throttling back to cruise power, I noticed that the fuel-
flow gauge indicated 14 gph instead of the usual 16. Since the
other engine instruments were giving normal readings, I concluded that the gauge simply needed adjustment. I intended to
tell the mechanics when I returned to Juneau.

My first stop was Tenakee, a hot-springs resort community of
about 80 people. I dropped off a passenger and a load of mail
and freight. While taxiing out, I had to give the engine a spurt
of fuel with the boost pump to keep it from dying—idle set too
lean. I'd mention that, too.

The fuel-flow gauge read just 12 gph during the climb-out
from Tenakee, and the needle slowly sank to zero. The rest of
the instruments continued to report healthy vital signs; obviously the fuel-flow gauge was broken. Occasionally the smell
of gasoline wafted my way, but I assumed the odor lingered
from a chain saw I had unloaded at Tenakee. As I worked my
way down the west coast of Admiralty Island toward the Tlingit
Indian village of Angoon, sleet and a gusty east wind began

thrashing the 206—the edge of a front forecast to arrive in the Juneau area later in the day. The wind, whipping the water into angry waves offshore, forced me to crab about 30 degrees inland to maintain a straight track. Fighting the gusts, I splashed down to a landing in the harbor at Angoon. The engine quit when I throttled back to taxi.

"Darn this idle," I mumbled to myself, restarting the engine. At the dock I unloaded my other two passengers and the remaining mail and freight. The company agent at Angoon later told the dispatcher in Juneau that he thought the engine sounded rough when I took off.

My last scheduled stop was on the east side of Admiralty Island, where a couple had radioed from their cabin for transportation to Juneau. I squinted inland, but a curtain of murky sleet and scud blocked the passes across the island. I would have to continue south down the shoreline and follow the coast all the way around. I was calculating my ETA at the cabin when I noticed that the airspeed needle was indicating only 90 knots, in spite of a cruise-power setting and an empty airplane. Was the airspeed indicator breaking down, too? According to my eyeball groundspeed check, the reefs and bights were passing more slowly than they should have, but then the wind was slightly on the nose and jolting the airplane harder now.

I checked the mags. Both were a little rough. So was the engine. It was rasping like a speaker who needed to clear his throat. And there was that gasoline smell, too. I pushed the prop and throttle controls to 25 squared—climb power—but the airspeed needle crept no higher than the 100-knot mark, still 15 knots below cruise. For another minute I tried to convince myself that the low airspeed was a result of my holding a nose-high attitude to counter the downdrafts, and that the engine roughness was merely a combination of rattling from the turbulence and whistling from the air leaks in the weary airframe. But wishful thinking could not mask the growing evidence that something bad was going on up front.

With a curse I banked away from the shoreline to begin a turn back to Angoon. As I glanced out the right window to examine the scuddy horizon, I noticed liquid sloshing on the floor of the footwell in front of the right front seat. Knowing what I would find, I reached over, stuck my finger in and smelled. Fuel.

The specter of a sudden inferno sent adrenaline tingling through my veins and nearly choked me with claustrophobia. I was now a mere 10 miles south of Angoon on the west coast of Admiralty Island. But 10 miles suddenly seemed a world away. If the cockpit erupted in flames before I reached either the village or the shelter of a cove in between, the east wind would sweep me into the swells of the open strait after I ditched. No one lasts long in Alaska's waters without a survival suit. On the other side of the strait bordering Baranof Island—which I could easily reach in moments with the strong east wind at my tail— I'd at least drift toward shore if I had to ditch.

I pointed the nose west toward the outline of Baranof Island, barely visible through the falling sleet, and pushed the yoke forward until the floats almost licked the water so that I could plant the airplane and dive out, if necessary. As the 206 shot across the swells, I flicked off all the electrical switches. Then I pulled the fire extinguisher from its harness and placed it in my lap. I never thought to slip into one of the aircraft's life jackets.

At Baranof I turned north, hugging the waves and the rocky shoreline. I knew that the nearest settlement was back at Tenakee, 30 miles away. Should I try for it or land in the first protected cove? Search aircraft would scour Admiralty Island first. It might be several days before someone got around to this stretch of Baranof—if the deteriorating weather permitted searching at all. As I watched the waves smash into the rocks, spraying the trunks of the spruce and hemlock trees that cluttered the shore, I felt overwhelmingly lonely.

I hesitated, then pressed the master switch, cringing against an explosion. The engine rasped on. Next, I gingerly pressed the avionics master and cringed again. The fourth frequency that I tried connected with one of the mountaintop VHF repeater stations that the FAA has erected along the coast in recent years.

"Cessna Four-Five-Seven-Five Quebec, Sitka Radio. Go ahead."

The voice was like a puff of air to a suffocating man. I explained my situation, said I was going to try to make it to Tenakee and asked the specialist to notify the company. Although I really didn't want to, I kept an eye on the growing pool of fuel on the right-hand floor. The odor gradually made me nauseated and I opened all the air vents within reach. Before takeoff I had filled the left tank, leaving a quarter-full right tank as a reserve. Now the left gauge indicated only a quarter-tank remaining as well. Enough?

I began to talk to the engine, "Come on, come on. Keep going, keep going." Then, just before Basket Bay, I flew out of the sleet at last. The gusty wind persisted, but my spirits rose with the visibility and ceiling. As my chances of making Tenakee increased, so did my impatience to get there. I fidgeted, ground my teeth and stared at the panel clock, which had slowed to a snail's pace.

At last I rounded the southern corner of Tenakee Inlet and was able to turn west, gaining a tailwind. I lost contact with Sitka Radio because of the intervening mountains, but there were periodic cabins along the shore now, a trawler plying the water and, up ahead, the village itself. Despite the roiling chop of the water below, I decided against using the electric flap motor as I turned upwind for the approach. Moments later the airplane shuddered as the floats bounced heavily across the waves, then fell off the step.

The engine quit. I had to restart it twice during the taxi to the dock. The company's agent grabbed the wing strut and pulled the 206 to the bumpers. "Just can't stay away, huh?" he said. The air smelled deliciously sweet as we walked up the ramp, and I breathed deeply to clear my head. At the general store, I gulped a Coke to dilute the paste in my mouth. Then I phoned Juneau, asked the dispatcher to tell the FAA that I was safely on the ground and described 75Q's symptoms to a mechanic.

Sleet was coating the dock 45 minutes later when the second 206 taxied in. After removing the top half of 75Q's cowling, the mechanic needed only a couple of minutes to locate the gremlins. In their haste to correct the compression trouble, he and another mechanic had neglected to tighten two fuel lines. One served the fuel-flow gauge and was the source of the fuel leak into the cockpit. The fuel had gathered behind the firewall and spilled onto the floor through openings. Meanwhile, the second loose line to the number-five cylinder fuel injector had caused the power loss—and sprayed raw fuel continuously over the hot engine. The mechanic secured the lines and replaced the cowling, and I dumped 10 gallons of fuel into the left wing tank from jerry jugs. The mechanic accompanied me back to Juneau in case more problems developed.

There was no debate about whether I had been tardy in taking precautionary action. Bush pilots tend to dismiss malfunctions and discrepancies more readily than other pilots. A broken instrument, a cargo door that won't latch, a bad leak in a float compartment—what are these to one who flies by the seat of his pants in isolated wilderness and wet, windy weather?

A pilot who grounds an airplane or turns around because of routine squawks quickly earns a nickname up here—dude. In bush country you ignore squawks until the job is done. If you have to stretch a few rules along the way, well, this is Alaska, not the lower 48. Thus I had ignored clues to a worsening situation until it became an emergency.

"Jack," I yelled above the engine roar, "what could I have done to reduce the fire risk?"

He shook his head and grimaced. "I think you were pretty darned lucky," he said. "It should have burned."

*The mechanics had left two fuel-line connections loose. They very nearly ruined the bush pilot's whole day. In the next account, a nonpilot sitting in the cockpit of a Bell Kingcobra caused his son some embarrassment. It could have been far worse. It took four attempts to take off before the pilot finally figured it out and got it right.*

# Pedal Pushing
### by Ben L. Brown

It was the spring of 1944. I had just picked up a new Bell P-63A Kingcobra fighter at the Buffalo, New York, factory. I had to deliver this plane to the Russians at Fairbanks, Alaska. My aircraft was one of the first half-dozen off the assembly line and was a completely new fighter, a great improvement over the P-39 Airacobra.

I filed my flight plan; my first stop would be Madison, Wisconsin, for fuel. I would fly over Toledo, South Bend and Chicago, refuel at Madison, continue on to Fargo, North Dakota, and stay that night. Next day, I would fly nonstop to Great Falls, Montana, where the aircraft would be winterized, ultimately continuing on to Fairbanks.

The flight plan was too tempting. I'd spent many months away from home ferrying all types of aircraft; I decided to stop in Toledo for half an hour to visit my parents and relatives. This was not official at all, and if anything happened to the aircraft at Toledo, I would be in very deep trouble.

I landed at Toledo and let everyone see the aircraft. My father wanted to sit in the cockpit to see what it was like. I helped him in, got him seated, and gave him a little cockpit check, explaining the instruments and levers. While he was in the cockpit, I ran into the little airline terminal building for a quick trip to the potty. I came out and got ready to leave, and said my goodbyes to everyone.

I fired up, taxied out and prepared to take off on Runway 22. I knew the airport very well, since I began my private flight training there as a 16-year-old kid in 1938. After doing my cockpit checks, I lined up for takeoff. For the first few hundred feet, you have to use a lot of right rudder until the rudder takes over on the takeoff roll. I suddenly found myself going to the left very quickly in spite of using full right rudder.

Before I could correct, I went off the runway and into the grass, bouncing around all over the area. After getting the aircraft stopped, I sat there and tried to figure out what was wrong. So I taxied back to the runway and tried another takeoff. Same thing happened.

After four takeoff tries, I was dumbfounded: Why can't I hold it straight down the runway on takeoff? Why is it veering off to the left so quickly? I was really sweating by now; what would I do if I couldn't take off? What has suddenly happened to the aircraft to make it do this? I was scared of what would happen to me, because stopping in Toledo was against the rules. It seemed like my number was up.

I finally gathered my senses together and tried to figure out what was wrong. I climbed out of the aircraft as it sat there on the runway with the engine idling, and took a quick look at the rudder; I thought it might have come loose. But everything looked normal. I climbed back in the cockpit, and once again tried to think it out before I made another takeoff attempt.

I raised my head toward the sky, looking through the glass canopy overhead, maybe in prayer. Then, like a bolt of lightning, it hit me. I noticed in the little rearview mirror that the rudder was cocked to the left, when it should have been in the middle or neutral. I rechecked my rudder pedals and found that even if I pushed the right pedal all the way in to the right the rudder only went over to the right a little bit. I quickly figured out that my father had hit the rudder adjustment stop on the outside of the rudder pedal, and thus the rudder pedal had just moved forward all the way. When I got in and neutralized the rudder pedals, the rudder cocked to the left almost all the way it could travel. I quickly readjusted the right rudder pedal; I could now see that the rudder was in the middle, and I took off without any more problems.

This taught me several lessons: to remember my Murphy's Law, which says that things will always go wrong at the worst possible time; to always adjust my seat or rudder pedals before startup; and to be sure the aircraft is locked before I leave the ramp.

# 6

# When the Go Juice Is Gone

An engine failure is a sure way to get an airplane back on the ground. Many times when the engine stops there is nothing the pilot could have done to avoid it. Often, but not always.

It's surprising, reading through the National Transportation Safety Board accident reports, how often pilots simply run out of fuel. They plan a flight that's going to take them four and a half hours, take off with four hours of fuel on board, and then are surprised when fuel exhaustion silences the engine four hours into their flight.

But running out of fuel is not always simply a matter of forgetting to top the tanks, or miscalculating headwinds or fuel flows. The fuel system can also fail in various ways. The experiences related in this section are not simple fuel exhaustion cases. Invariably, there was another, uncontrollable factor, that worked its insidious magic to bring down the plane and pilot.

## A Fool and His Fuel
### by Johan Vejby

Throughout my flight training days in South Carolina, I had always taken great pride in my cross-country fuel-burn calculations; on occasion, I had figured fuel down to the nearest tenth of a gallon, with just the legal reserve remaining in the tanks. But I was able to do this with a degree of accuracy because I was familiar with the operation of the school's well-maintained airplanes, mostly Cessna 152s.

Shortly after graduation in May 1987, I began working as a flight instructor. During this time I tried to be mindful of what an instructor of mine once told me: *Your flight is as good as your preflight.* I was soon to learn the importance of his advice.

Then I got my first "professional" assignment: My boss asked me to fly a Cessna 150 from the small Alabama airport where I was working to Memphis to pick up his son. I had never flown a 150 before, but a few swings around the pattern before embarking on the three-hour round trip convinced me it was just like flying the familiar 152.

For preflight I glanced through the POH and got a weather briefing. The forecast called for late afternoon scattered thunderstorms over Louisiana. Then I scribbled some notes regarding fuel consumption and intended course—actually, I wrote down only the names of a couple of VOR radials to follow and the estimated time I would reach reserve level. This was my first professional flight, so I stretched reserve time to one hour. Acting the role of "professional pilot" I deliberately set up a firm departure time in order to impress my boss with my efficient planning ability.

I departed on time and flew direct to Memphis without problems. There I met my boss's son, and following a brief preflight and weight-and-balance calculation, we were Alabama-bound. The flight progressed, and we were late reaching our first checkpoint. Wanting to check the sectional, I realized that with brilliant cockpit organization, I had stowed it in my flight bag—which was back in the baggage compartment.

My boss's son helped to extricate the map from my bag. I discovered that we were almost 30 miles off course and that something was amiss with the CDI. The same instructor who had advised me about the importance of preflight had also insisted on my being aware of and pointing out my position on the chart every five minutes, regardless of whether the CDI was centered. "The navaids are there to help you with the map," he'd say, "not vice versa."

But things were not that bad. . . yet. I used pilotage and plotted a rule-of-thumb course, proceeding like nothing happened so as not to worry my passenger. The weather appeared to be worsening, and I soon found it necessary to circumnavigate a couple of cumulonimbus clouds. Keeping my passenger diverted, I explained the weather phenomena and permitted him to handle the controls for a while. At this time, I noted that the fuel gauges were indicating lower amounts than I had expected, but as the flight went along, they seemed to stabilize and agree with my calculations.

Crossing the border between Alabama and Louisiana called for a change of sectionals—but I realized the sectional I needed was back in my office. Fortunately, I recognized an airport to the east and vaguely remembered I would have to fly a direct northerly heading, past another airport, to reach home port.

As the adrenaline pumped through my body, I became more critical of my "professional" skills, which included having failed to lean the mixture. Then, with E-6B in hand, I tried to compute just how much fuel had been unnecessarily consumed as a result of my failure to lean, and how much fuel was remaining. That old instructor of mine had also warned me: "Never fly on your reserve unless you are having an emergency." We had 45 minutes of flying time left. At this point, my passenger was eyeing me with less than complete confidence. When I explained we might have to land and refuel before we got down to reserve, which I preferred not to dip into, my passenger was calmed somewhat—although the situation wasn't improved by my asking to borrow money to pay for that fuel.

Fifteen minutes later, I spotted an airport to the east. The fuel gauges showed quarter tanks remaining, but the gauges had indicated this same amount for quite some time. Even though I had calculated it was only 10 minutes to home port, I did not want to go to reserve fuel. In addition, I was too low to get radar service for direct vectoring, so I decided to go to the only airport in sight.

But then I recognized the familiar hill near the home airport and, disregarding my instincts to land at the closer airport, went for it. I set up for a long final and began a descent, estimating we would cross the hill at about 1,000 feet agl. Now I felt calm, and I knew we would make it. There was still plenty of reserve left, was there not? Four minutes later, the engine answered the question with a cough, then silence. I fishtailed the airplane and it coughed twice more before it shut down again. In spite of gauges indicating quarter tanks remaining, I knew I had run out of fuel—in the vicinity of three airports.

This was an emergency. I looked for other places to land, while running through the emergency checklist. Located behind us, a small road—about 1,000 feet in length—was to be our runway. Established on final, and right on speed, I stared at the motionless prop blade and realized this was not a training procedure. I only had one chance.

A boy was riding his bicycle on the road, probably amused to see an airplane up close. I lowered the nose to pick up speed, then raised it again to be sure I landed beyond him. But I lost a few hundred feet of runway in the process, only to discover the road went uphill before it leveled off, overshoot another 200 feet. Suddenly, my 1,000-foot runway was only 500 feet long.

I flared just before ground effect and settled firmly on the mains. With shrieking tires, I came to rest within a hundred feet. My passenger was smiling and then he danced around the airplane. Stepping out of the airplane, I sat down on the footrest, because my knees just would not carry me. There were four inches of clearance between the wingtips and the trees along the road.

A farmer helped us pull the airplane off the road. Later, we got some fuel into the tanks, the police cleared the road, and I took off to bring her safely home. We learned the airplane had used almost 37 percent more fuel an hour than listed in the POH, even with full-rich mixture taken into consideration.

I kept my job as a result of "skillful flying." Now when I instruct, I emphasize the importance of thorough preflighting and knowing emergency procedures. I never got over that incident, and never will.

✈ ✈ ✈

*Proud of his fuel-burn calculations, the pilot, unfamiliar with the Cessna 150, was nevertheless convinced he still had at least half an hour of fuel in his tanks. An alternate airport beckoned him, but he was seduced by the closeness of his home port and elected to continue on. Not long after that, things got real quiet.*

*In the next event, the pilot of a Musketeer learned that the full tank he was selecting wasn't full and that fuel still in the pumps on the ground doesn't help keep the engine running.*

# Forced Field
## *by David Kraft*

The weather was ideal. I arrived at Delano Municipal Airport, Delano, California, for my scheduled student solo practice eager for the thrill of "pilot in command" flight.

Quickly I went through the preflight on the Musketeer. A little too quickly, as it turned out. While checking out the cockpit, I noticed that the fuel-selector knob was completely detached from the valve on the bulkhead. With my mind on what I would be doing once aloft, I casually replaced the knob, placing little importance on its correct alignment on the valve. That was mistake number one. A quick peek inside both fuel tanks revealed the left tank to be practically full while the right was almost empty. No problem. I'm only going up for an hour or so, and besides, this is a local flight. What could possibly happen? Mistake number two.

My mission for the day included practice on ground-reference maneuvers followed by 30 minutes of touch-and-goes. The takeoff and the short hop to the practice area east of Delano were pleasant and uneventful, and the airwork went off without a hitch. I congratulated myself for my obvious mastery of the airplane, and looked forward to showing off my finesse in the landing pattern. My instructor would surely be watching me from the ramp; it had been barely two months since my first solo. My 26 hours of flight time notwithstanding, I fancied myself quite the seasoned pro as I pulled the power all the way back, then flared. The subtle, satisfying squeak of the tires on concrete signaled a textbook landing.

As I took off again, I was already planning my next landing. Suddenly at 400 feet my euphoria was interrupted by a threatening shudder coming from the nose of the airplane. In horror I watched as the propeller wound down to a halt. I had run the near-empty fuel tank dry, thinking all the time that the engine was feeding off the full tank.

I froze—for a few fleeting moments I was no longer the pilot in command, but merely a helpless passenger without the means to alter a cruel fate. This just could not be happening to me.

The departure end of Runway 32 pointed straight to a moderately wooded residential area liberally interspersed with telephone poles and electrical lines along narrow streets. Where could I put down without killing myself or some unsuspecting soul on the ground?

Just as I was bracing myself for the inevitable, something caught my attention off to the left. Across the street from the airport I spotted the choicest piece of real estate I had ever laid eyes on—a long, flat, vacant lot, situated right next to, of all places, a funeral parlor. With hands and feet flailing all over the cockpit I yanked down the flaps and set up an abbreviated approach for that oasis.

The mains touched down hard, but not too hard. I stood on the brakes, and the Musketeer came to rest a little more than halfway down the length of the vacant lot. The instant the airplane stopped, I jumped from the cockpit on wobbly legs. In a daze, I looked the airplane over and realized I was still alive.

Minutes later I was joined on the scene by about 20 vehicles speeding toward me with tires screeching and clouds of dust flying. My instructor turned up and was pleased that both student and airplane had emerged unscathed. We boarded the aircraft and, after switching the fuel valve to the full tank, started up and taxied down the road to the airport entrance. I had had enough flying for one day.

When I got to the office, I was led by the arm to a small room where two men from the FAA were waiting. Upon their request, I related in detail all the events. Then they shook my hand and congratulated me on the successful outcome of the incident.

But there was little worthy of praise in my performance. Mistakes were piled one on top of the other. First, there was the improper placement of the fuel valve and my failure to grasp its importance. Then there was my decision to fly with one empty fuel tank. It was pure chance that the engine quit at the right moment to place me in a position to successfully land with so little time and altitude to spare. All the ingredients for a classic stall-spin accident were there in more than ample amounts. My impatience to get into the air that day clouded both my judgment and common sense.

*Misaligning the fuel selector handle didn't help the Musketeer pilot's case, and he did make some questionable decisions. Nevertheless, he was aware of how much fuel there was on board and believed it was feeding from the full tank. Unfortunately, he was wrong.*

*The pilot of a Beech Sundowner also did everything right—almost—during his preflight. He just didn't understand what he was seeing when he checked for water in his fuel. Few of us would have.*

# The Longest Minute
### by Don White

After leveling at 2,000 feet, I received immediate clearance to 6,000. I looked over at my wife, who's a reluctant passenger at best, and wished she could share my marvel at the sheer joy of flying. Then, as I moved the throttle smoothly forward, the impossible happened. What moments ago had been a soothing, rhythmic heartbeat coming from the engine had now become the worst sound that can come from a single-engine airplane—nothing but the whir of the wind.

I'm an instrument-rated private pilot who flies only for pleasure, and I've been the proud owner of an 11-year-old Beech Sundowner for about a year. My wife and I had truly looked forward to this vacation for months, but from the beginning, the odds seemed to be against us.

Just getting my wife into the airplane is a major accomplishment, but getting her in with an 800-foot ceiling and one-and-a-half-miles visibility in fog was definitely a feat worth recording in the logbook. I didn't bother to mention the line of thunderstorms building off the coast. Besides, this was only a two-hour flight over familiar terrain, and the storms were forecast to be out of our destination area by the time we got there.

We climbed out of the fog and haze typical of Southern airports in the middle of summer; the first hour was fairly calm as we cruised in and out of the clouds. But we became increasingly concerned about the amount of cloud buildup and the snatches of weather updates that I was receiving from Flight Watch.

The thunderstorm activity on the coast had built into level-four and level-five monsters that engulfed our destination. When I began requesting deviations around ominous-looking cumulonimbus, my wife recommended that we turn back.

"No problem," I said, putting on my best Joe Cool face. "We'll just get close and if it looks too bad, we'll just land and wait it out." And that's exactly what we did. We waited out the storms, took off when they passed, and eventually landed at our destination under beautiful, sunny skies. We were totally prepared to enjoy a week of fun in the sun. Little did we know that was pretty much the last time we would see it for the entire trip.

It was the very next day that Tropical Storm Chris decided that it would also like to spend some time at our beach. It proceeded to visit with 50-knot winds and about five inches of rain. When the storm finally passed, I took a cursory look at my airplane and found no indications of anything out of the ordinary.

We got an early start the morning we left to avoid any afternoon thunderstorm activity. The sky was 2,500 feet scattered, occasionally broken, with five miles visibility. I was concerned about the potential for water in the fuel because of the rain that had fallen during the week, having no faith in the integrity of the recessed fuel caps on my Sundowner. I was surprised to find, however, that no water came from any of the three sump drains. So, full of confidence, I copied our clearance and took off, anxious to get home.

We were exactly five miles from the airport when the engine quit. "What's wrong?" asked my wife. As my left hand squeezed new finger impressions into the yoke, I replied with an inspiring, "I have no idea," and proceeded to establish best glide, switch tanks, check the fuel boost and mags, talk to ATC, look for a place to land and perform 573 other items just in case. I received vectors back to the airport but quickly realized that we would end up about a mile short of the threshold. I lined up for approach on a small highway, the only patch of solid ground that I could see besides marshland.

As I concentrated on getting down safely, I had the nagging feeling that water in the fuel was causing my problem, so I tried rocking the wings and making significant pitch changes, which caused the stall warning horn to sound. For the moment none of these maneuvers seemed to be having a positive effect on the engine, not to mention my marriage. But although it seemed like a month and a half since the engine had conked out, it was hardly a minute before the miracle occurred—the engine cranked back up. With the prop still windmilling, we were back in business.

I climbed to 3,000 feet, and ATC asked if I wanted to resume my flight plan. My wife quickly said, "Get us back on the ground now." So I informed ATC that we would be returning to our point of departure and that the emergency condition had ceased to exist—when the engine promptly died again.

This time, however, it wasn't a quick death, but a lingering illness with oscillations ranging between 500 and 1,500 rpm. I was able to return to the field, high and hot, and made a reasonable facsimile of a carrier landing.

We taxied to the ramp with the engine still surging. As I was getting out of the airplane the line personnel rushed over to make sure everything was okay. Once they realized we were both fine, one was quick to say, "Well, don't feel bad, an astronaut landed here yesterday with the wheels up."

With that image to comfort me, I tried to reconstruct what had just happened. First I checked the airplane's fuel, where I found copious amounts of water in each tank. This confused me. How come I didn't pick this up on my fuel inspection prior to takeoff?

Apparently I saw no layer of blue gas and water because I was looking at 100 percent water through a sky blue background. I was also fooled by the presence of a residual odor of avgas. I can remember looking at my fuel samples and thinking that they seemed a very light blue. Now I use my white airplane as background when checking fuel, and then check for puddling when I pour the sample on the ground. I have also started taking several samples from each drain, instead of just one.

On my arrival I had asked that my tanks be topped off, but I know for a fact that they were not filled until the day of my departure. The half-empty tanks allowed almost two quarts of water into the fuel system. But, of course, if the O-rings on my recessed fuel caps had not been cracked, no water could have possibly leaked in.

In retrospect, I'm a much better pilot as a result of my experience. I have learned to pay attention to my gut instinct. It's been said that success relies on a combination of intuition and logic. In this incident, I should have paid more attention to both.

✈ ✈ ✈

*There was fuel in his tanks, but there was also water; much more than he realized when he did his fuel sampling. If it looks like water, feels like water, and acts like water, it probably is.*

*In the following incident, the pilot of a Cessna 150 fueled his own airplane and knew he had four hours on board. Nevertheless, the engine stopped for lack of fuel three hours and 15 minutes later. Had he run out of fuel? What else could it have been?*

# Vent Frustration
## by Alan Gjerde

I departed Emmetsburg, Iowa, on a bright, cool morning in mid-April. My mission: to deliver my old Cessna 150 to its new owner in Baltimore. I had topped off the tanks myself before I left, so I knew that I had four hours of fuel. After checking my groundspeed again and again (what else is there to do on a long cross-country in a 150?), I determined that my first fuel stop would be a small, uncontrolled field on the south side of Chicago. The trip would take a grand total of three hours and 20 minutes.

After grinding along for three hours and 15 minutes, the engine stopped, without so much as a miss to indicate that there was any trouble. I was within two miles of the airport, but it might as well have been 50. All I saw below me were trees and a single highway. I would not have believed anyone who told me that there is a national forest inside the city limits of south Chicago, but I was over it.

After a frantic effort to find out why the engine had quit and an attempt to restart it, I lined up with the narrow highway that was underneath me. My timing could not have been worse—it was now 9:15 a.m., and highway traffic was heavy. I stayed lined up with the highway as long as possible, but there was no lull in the traffic. I pulled off to the right and started to slow the 150 to a stall just as it hit the top of the 50-foot trees. The tail and the left wing hit first, and the airplane pulled sharply to the left and pointed nose down. Suddenly I realized that I might get hurt. When the airplane started to fall through the trees the wings hit progressively larger limbs, breaking the fall. When I finally hit the ground, I hit tail-first with little damage to the cockpit area.

I had shut the master off before the crash, so all that remained to be done was to collect my suitcase, my maps and my rubber legs, and get out. I dropped about five feet to the ground and walked about 20 yards to the highway, which was now deserted. After what seemed like an eternity, someone stopped and asked what had happened. I was just as curious as they were. Even though I was the first person to arrive at the scene of the accident, I didn't know what had caused it. When the engine quit my prime suspect was fuel exhaustion, but I looked at the gauges and the left tank registered one-fourth full; I moved the fuel selector on and off and moved the mixture to full rich, all with no results. When the FAA arrived, I was grateful to talk to someone who could help me find out what had happened.

We tried to take samples from both wing drains, but there seemed to be no fuel. That was the determination. I was angry, embarrassed and disgusted, and just wanted to walk away from the whole situation; but I had an airplane sitting in a national forest and the authorities wanted it out of there before nightfall. So I called a salvage yard and they came and hauled my 150 away.

Three days later I received a phone call from someone who had gone to the salvage yard looking for a fuel cap for his 150.

"You know," he said, "when I took off the fuel cap from the left wing of your airplane, there was a 'whump' sound and about three or four gallons of gas ran out of the severed fuel line." He also told me that when he checked the vent on the cap, he could get no air through it. The mystery had been solved.

After determining the length of time that I had flown, and also that I had filled the fuel tanks, the FAA agreed with me that the probable cause of the accident was a vacuum created by a plugged fuel-cap vent.

It happened to me, and it could happen to you. Check the fuel vent to make sure it's doing its job. If you hear air rushing in when you take off the caps to fill the tanks, your vents may not be working.

✈ ✈ ✈

*The blocked vent on the Cessna 150 prevented the fuel that was still in the tanks from feeding the engine. In the next account, the pilot of another 150 also has fuel in his tanks. But again, the engine isn't happy with what it's getting.*

# Tiny Bubbles
## by Ronald Wanttaja

I keep my Cessna 150 at an airport near work, to take advantage of any breaks in the ever-present Pacific Northwest drizzle. On one such break I took off for a quick lunchtime hop. Since I was down to a quarter tank of fuel, I decided to stop at a field where I used to base my airplane. On final, the airspeed indicator became erratic, apparently due to a pitot tube waterlogged from recent heavy rains. It had happened before and was nothing more than a minor annoyance. Nevertheless, I taxied to the A&P's hangar. Like me, he was on a lunch break, but at least the gas pumps were open. It was still before noon and I was only the third customer of the day.

When I'd first learned to fly, in the Midwest, checking the fuel for moisture meant a quick spurt from the sumps. But I'd never been completely comfortable with that, and since coming west I had begun using a sampler to check the fuel before flight and after refueling. So after topping off my tanks today, I took a sample.

There were a number of tiny bubbles in the fuel. I knew what water looked like. The shimmering silver of condensation in a fuel sampler is a common sight in the damp Northwest. But bubbles? I took another sample. Same things. The right tank was the same way. I carried my sample to the line shack.

The crowd was interested. "Filling up with champagne?" someone asked.

"It certainly costs like it," I replied.

Guesses included dirt stuck to the inside of the sample cup and air mixed with the fuel due to the cavitation of the pump.

"Couldn't be water," someone mused. "The pump's got detectors and valves that shut off the flow if there's water in the gas." I took another sample. Still bubbly.

I fired up and taxied to the runway. Instead of flying the few miles back to my home field at pattern altitude, I blasted on up to 2,000 feet. Looking for a forced landing spot would be good practice, and I started examining the countryside. Three miles from home, the engine missed slightly. I sat up straight. It missed again.

I turned on the carb heat, switched mags and fiddled with the throttle, but I knew what it was—the bubbles. I pushed the throttle to gain as much height as possible. The roughness increased, as did the missing and the catch of anxiety that was growing in my throat. I was approaching my home field from the south, with traffic landing to the south. I could see a 152 just turning final and a Tomahawk on base, so I announced, "I've got a little bit of an engine problem here. I'd appreciate it if you folks cleared the runway down there."

Even though the propeller stopped dead on final, the landing was "normal" and I turned off at the centerpoint, rolling toward the gas pumps. I took a sample from the left sump drain. The bubbles were gone—replaced by filmy swamp water. There was not a trace of red-tinged 80-octane. I grabbed a telephone and warned the FBO to check its system.

In all, it took me two hours to drain most of the water out of the fuel. I estimate that my 13-gallon fuel tanks each contained at least two gallons of water. The FBO called back and confirmed that its storage tanks were contaminated. The filling points to the tanks are underground, recessed in small manholes. The heavy rains had filled the manholes and a leaky cap had allowed water to enter the tank. The water shutoff valve was also faulty, only slowing the flow instead of stopping it completely. Water had mixed with the avgas while it was being pumped, causing the bubbles, and then had separated from the gas slowly, thus allowing start-up and takeoff.

My bubbly fuel sample had been examined by a panel of experts that had included everyone from the FBO to a CFI. No one had suspected water. We had all assumed that the automatic shutoff would ensure the purity of the fuel. It had been a four-way failure that nearly wrecked an airplane—recessed fill points, leaky caps, a faulty valve and a pilot who looked right at the problem and didn't recognize it.

What of the two airplanes that had filled up before me? The first had flown to two other airports before the FBO contacted him; the pilot drained less than a cup of water. The second airplane, which had fueled minutes before I did, was based at the field. The pilot had landed, refueled, parked and gone home. Good thing. They drained more than five gallons of water from his tanks.

*The pilot did everything he thought he should. He sought out what he thought was experienced and professional counsel and acted on it. Unfortunately it wasn't good advice.*

*We've all been taught to visually confirm the amount of fuel in the tanks and the pilot in the next account did just as he was instructed. How was he to know that what looked to him like full tanks weren't?*

# Fuel Crisis
### by Brian Lloyd

A couple of months past my nineteenth birthday I found myself a student pilot for the second time. I was planning to take my commercial check ride and was a bit short on required cross-country dual IFR time, so my instructor and I planned a flight from Brackett Field in La Verne, California, to San Diego's Lindbergh Field and back again, under the hood. The trip would take about three hours of flight time.

The preflight seemed routine to me and the aircraft had all its pieces in the correct places. Using a ladder, I climbed up to check the fuel in the tanks. The level was down from the filler neck about half an inch. "Full," I said to myself. I then checked the log and noted that no one had used the airplane all day. My instructor strolled out to the aircraft and I assured him that we were ready.

We terminated the first flight of the day by shooting the localizer back-course approach to Lindbergh. After touchdown we taxied in. I decided since we were near a gas truck, we should top the tanks.

"Let's put in some fuel," I said.

"No," my instructor replied, "The 150 has four hours' endurance and we've only flown for 1.5 hours. It was full wasn't it?"

"Sure, but the truck is right there."

"Don't worry," he said.

I filed IFR for our return flight and chose the coast route. The fuel gauges were looking a bit low as I prepared for the VOR-A approach to Brackett. ATC gave us some last-minute vectors for traffic spacing and then cleared us for the approach.

The VOR-A to Brackett is a circling approach with a very steep descent gradient. The course is 90 degrees to the runway and terminates at the Pomona VOR, which sits right underneath the midfield point of the downwind leg. I flew with the throttle closed so I could get down to the minimum descent altitude as I crossed the VOR. This would put me below the normal traffic pattern altitude. My approach went well, if I do say so myself, and I turned to enter downwind as soon as I crossed the VOR. When I opened the throttle to arrest my descent, nothing happened.

The airplane was abeam the runway threshold, so I began a turn to the airport. My instructor grabbed the yoke and turned us back onto the downwind course, yelling that I was number three for landing. He got the message when I pointed to the firewalled throttle. I turned back to the airport while he notified the tower of our predicament.

After we had landed, the engine caught as we turned off the runway; but about halfway to the ramp it quit again and nothing would bring it back to life. We pushed the airplane the last hundred yards.

We called for the gas truck and waited while the lineman fueled the aircraft. You could have heard a pin drop when he shut off the nozzle. The flow gauge showed that we had taken more than 24 gallons. The 150 holds 22.5 gallons of usable fuel and has a total capacity of 24 gallons.

My instructor berated me for not checking the fuel. I held my thumb and forefinger about half an inch apart: "It was only down about this much. I thought it was full!"

"Full?" he exploded. "It was down a quarter of a tank. Don't you know any more about a 150 than that?"

"I learned to fly in a 182. I've only flown 150s for local dual instruction," I replied.

Later we discovered that someone had flown the aircraft for about 45 minutes that morning and neglected to log the time. The dribble of fuel that remained had unported during the steep descent. Upon landing, that little bit was spent taxiing. We had burned even the fumes.

Even so, it could have been much worse. ATC vectored us for 10 minutes before clearing us for the approach. What if it had been 15 minutes? There are few places to land off-airport in Southern California. I am glad and lucky that I didn't have to try to find one.

# 7

# I Should
# Have Known Better

There's an old adage that says, "I'd rather be on the ground wishing I was in the air, then in the air, wishing I was on the ground." Pilots who take off for a flight when they know they shouldn't are asking for trouble. Many times they get it.

In each of the incidents related in this chapter, the pilots knew before they launched that there were good reasons why they shouldn't. They had a "feeling," or they knew better. It's almost as if someone was riding on their shoulder whispering in their ears.

Some of the accounts are by pilots who, faced with enormous pressures to make a flight, instead made the difficult decision to cancel. The decisions are obviously correct, but that doesn't make the process any easier. Pilots are often forced to balance the risks against the value of the trip—getting home, earning money, or saving lives. Accident summaries are replete with accounts of people who, because they were involved in life-saving flights, pushed themselves and their aircraft beyond their capabilities. Not only didn't they save any lives, they became accident statistics themselves.

After a very long day, a pilot finds himself at the controls of a Cessna 185 at night. He should have known better.

## No Time for a Nap
### by James F. Robak
I had just finished my afternoon rush-hour traffic reports for a radio station in Milwaukee, and was bringing our Hughes 300

helicopter in for the night. I felt the usual TGIF as the chopper settled softly onto the apron near the hangar. The thought of sleeping late on Saturday was foremost on my mind. I was tired from a very hectic week, teaching a full load of high school classes and flying six shifts of traffic reports, but I loved what I was doing and always found the energy and the time whenever there was a chance to do some flying. I would soon learn a hard lesson about pushing oneself beyond reasonable limits.

As I pulled mixture and turned off the master on the Hughes, the main rotors slowly rotated to a stop. While I was walking to my locker in the hangar, the dispatcher for the charter department summoned me to the phone. The chief pilot was on the other end; he needed someone to fly to Buffalo, New York, with him in order to pick up an airplane that was needed for a charter in the morning. None of the regular charter drivers were available. The weariness I felt disappeared when I thought about this cross-country trip. I said yes, I would do it.

While I was finishing a second cup of coffee, the chief pilot walked in and explained that we would take the Cessna 185 Skywagon to Buffalo where he would pick up the charter airplane and I would fly back in the 185. I was now feeling a second wind—or maybe the caffeine—and agreed that I'd rest on the way out while he did the flying.

I should have slept, but I hadn't seen the chief pilot for a while; and so we settled into conversation. As we taxied in at Buffalo I realized that I had not slept at all during the flight. Passing on the last sandwich in a servo-matic machine, I opted for another cup of coffee and some cookies. The chief pilot said that he'd be a while and that I should start back in the 185 as soon as I was ready.

The lineman had just finished topping off the tanks as I completed my walk-around and climbed into the left seat of the roomy Skywagon. It was a little after 10 p.m. when I nudged the wheel back on the big powerful Cessna and pulled up over the twinkling lights of Buffalo, taking up a heading of 270 degrees back to Milwaukee. On the climb to 8,500 feet, visibility was unlimited and I found the smooth cool night air very refreshing. The flight started out as perfectly routine, passing to the north of Lake Erie through Canadian airspace to the London VOR. From Flint it was just 100 nautical miles to Muskegon and the east shoreline of Lake Michigan, with only another 95 miles to Milwaukee.

Over the last 100 miles I had climbed to 10,500 feet for some insurance altitude and decided to go straight across Lake Michigan rather than head south along the shoreline and through the tangled web of Chicago's TCA. At this altitude I could already see a familiar light from Milwaukee all the way across Lake Michigan: the very bright light coming from the largest four-face clock in the world, atop the Allan Bradley building about five miles north of the airport. Visually it seemed so close; but it was still 90 miles from my position east of Muskegon as I crossed the east shoreline of Lake Michigan into pitch blackness except for my bright beacon on the other shore.

It was then that the accumulated events of the week and of that particular day conspired to create a situation I would never forget. I was physically and mentally tired from teaching all day and reporting traffic both early morning and afternoon. I had taken this flight without proper rest and proper nutrition. When you're over-tired, even several cups of coffee will only keep you alert for a short time. That, in combination with my flying at an oxygen-thin altitude, the soothing drone of the engine, the music I had tuned in on the ADF and my fixation on the ever-brightening light across the darkness, put me into a relaxed trance . . . and in a little while . . . I FELL ASLEEP!

A little bump knocked my head against the window and jolted me awake. I was startled; but I thought that it was just one of those moments that lasts only a fraction of a second, like when you nod off momentarily at your desk. I immediately opened a vent for some cold outside air to wake me up, and looked for my bright light. It was gone; instead of one beacon in the darkness, there were bright city lights all around. Where was I?

The theme from *Twilight Zone* began playing in my head. I was still disoriented, still waking up. The altimeter read 13,400; I had climbed 3,000 feet. Did I make some gradual turn that put me back in Michigan? Could I have drifted south into Indiana? The directional gyro read 265 degrees; I was only five off my original course, but I didn't recognize anything on the ground. I reduced power and turned toward the larger city to my right. Only then did I think. . . *FUEL?* . . . *Oh, my God, how long was I out?*

The fuel gauges showed needles bobbing just above empty. I had used up most of my fuel, and I was lost. I realized I had to get down right away or chance a forced landing in the dark in unknown terrain. As I approached the city to my right I noticed a large lake in the center of the metropolis and a large, white, brightly lit building that looked suspiciously like our capital. I realized that I was over Madison, Wisconsin. I had overflown Milwaukee by 100 miles, or about 45 minutes. What I had perceived as just nodding off for a second was actually falling asleep at the controls for one hour and 25 minutes. I had flown 200 miles, solo cross-country at night and over Lake Michigan, fast asleep. I was very lucky—and thankful that I was still alive.

I called the Truax Madison Tower and reported myself as 10 miles southwest. A friendly voice told me to report having the field in sight and that I was cleared to land. In a matter of minutes I had touched down, taxied to the nearest FBO, found some tie-down space on the apron and put my big tired friend to bed—and then, after a quick and chagrined call to Milwaukee, did the same for myself, at the nearest airport motel.

✈ ✈ ✈

*What if he hadn't been jolted awake when he was? Would he have flown along until he ran out of fuel and then crashed? He knew he was tired, knew he wasn't at his best. He should have known not to make the flight.*

*Here's a pilot whose friend knows better and suggests the pilot leave his airplane and ride home in a car. "How long does it take to climb to 300 feet?" the pilot asks himself. If he didn't know better before, he did after this flight.*

# Ice Capade

### by J. D. Creley

In 1973, at the age of 31 with time on my hands and my finances adequate, I decided to fulfill my lifelong desire to learn to fly. After checking with several flight schools, I chose one at Merrill Field in Anchorage, Alaska. This school had the highest student success rate in Alaska. My flight instructor was one of the best in the state. After 40 hours of instruction and solo flying I passed my private flight test and began logging as many hours as possible.

After months of searching I found and purchased my first airplane. The 1968 Cessna 150 gave me many a happy hour of cross-country flying. I made numerous trips to other airports in all kinds of weather, and I learned about the hazards of winter flying in Alaska. But one hair-raising experience, on a wintry day in February 1974, still burns vividly in my memory.

I was working on a winter construction project about 45 road miles from Anchorage, Alaska. Because of the heavy rush-hour traffic to and from the military bases, this 45-mile drive took about 90 minutes one way. I had discovered an abandoned automobile racetrack close to my work area and, realizing it would make a perfect airstrip, I obtained permission to use it. I began flying my 150 to work. With only a five-minute drive to the airport and 15 to 20 minutes' flying time, I could make the round trip in about one-third the time it took to drive. Flying to work soon became routine.

The forecast for this long-remembered day was for cold, clear skies and patchy fog. I foresaw no problems, but when I arrived at my usual landing strip it was completely obscured in ice fog. I didn't have time to return to Merrill Field and drive to work, so I decided to try landing on the truck haul road at my work site. Flying down the river that skirted our work area, I descended to about 100 feet agl, at which point I could see the haul road below the fog bank. I flew under the ice fog for about half a mile and landed. By the time I had tied down the 150 and covered the wings, the ceiling and visibility were zero-zero.

During my entire eight-hour shift the ceiling and visibility did not improve. I discussed with my friend, a co-worker, the probable thickness of the fog. My friend, a pilot with many hundreds of hours of experience, suggested the fog could only be 200 or 300 feet thick. He also suggested that I leave my airplane and ride home with him in his car. But I feared that my airplane would be vandalized if left overnight, and I knew that I could handle a short trip on instruments. "How long does it take to climb to 300 feet?" I asked myself.

After work I removed the wing covers and cowl cover and started the 150's engine to warm it up. I had asked my friend to drive his car to the end of the haul road and aim his headlights toward me as guidance for a straight takeoff run. I taxied to the haul road, completed my takeoff check and set my altimeter to 100 feet. I could easily make out my friend's headlights at the end of the road. I turned on the pitot heat, shut off the carburetor heat and came in with the power.

My overconfidence continued until I lifted off, at which point I suddenly realized I had made the biggest mistake of my life. I knew that ahead of me, buried in the fog, were an icy river and trees 40 to 50 feet high. I also knew that my only option was to continue. I had experienced vertigo with the security of an instructor sitting by my side, but now I was alone. For the first time I knew what real, life-threatening vertigo was like.

The fear that gripped me at that moment is unimaginable. A quick scan of my instruments showed I was in a right turn. My thoughts of the trees now ahead forced me to level the wings, and the thought dawned on me that my life was literally in my hands.

I began to shout at myself in an attempt to overcome the vertigo. By this time I was climbing through 400 feet indicated, with no break in the fog. I still don't know if the shaking of my knees was caused by the cold or by the cold fear inside me. I was glad I'd had the presence of mind to tune the Big Lake VOR before takeoff. If I maintained a heading toward it, I would not crash into the mountain that loomed to my left.

Climbing through 700 feet I noticed the windshield was covered with rime ice. I pulled the carburetor heat on, and the engine coughed as it inhaled the ice particles that had formed in the carburetor. A small area of the side window remained ice free, and by bending forward and looking out I could see that I was skimming the top of the fog bank and that my wing strut was covered with ice. My altimeter indicated 1,150 feet. I had climbed through more than 1,000 feet of ice fog that I had supposed to be 300 feet. With a second look I noticed that the fog was now behind me, and I could see Cook Inlet below.

The ice that had collected on the propeller began to break off, causing tremendous vibration. I reduced the power and the shaking subsided. For the next few minutes I was busy trying to keep the wings level, as large chunks of ice broke off the wings, first right, then left and right again. I flew around the area for about an hour and a half, until the ice on the airplane had dissipated enough to return to Merrill Field.

After regaining my composure I landed and immediately made arrangements to begin my commercial flight training. I now hold a commercial certificate and have greatly improved my headwork and attitude toward flying.

Thinking back over this experience, I am thankful I was flying a very forgiving airplane. I now realize that private pilots get only enough instrument training to get them into trouble, and that a private license is only a license to learn.

✈ ✈ ✈

*Lessons learned the hard way are often lessons that are well learned. The Cessna 150 pilot is not likely to lift off into ice fog again, no matter how thin a layer he thinks it is.*

*A candidate for his instrument rating, faced with a go/no-go decision on the flight test, learns a lesson that should go a long way to keeping him—and the rest of us—safe.*

# License to Learn

### by Tim Kenny

Our home airport was obscured by a gray mass the size of Rhode Island. ATC called it "intense thunderstorm activity." I swallowed the lump in my throat and turned to the FAA designated examiner in the right seat.

"What do you want me to do now?" I asked.

The examiner continued taking notes. "I'm just an observer," he said without looking up. "You're the pilot."

For the first time since I began my instrument training I was faced with an inflight weather decision. With all I'd been through in preparation for the check ride, I'd never anticipated this.

The two-hour oral had gone well, the ILS and VOR approaches without a hitch, and my NDB and holding work had been passable. Now I had to decide what to do about the unexpected turn of events in the sky above our destination.

I thought back to the security of my student-pilot days, when things were so much easier. My instructor signed off every trip. Together we'd review the weather picture, look at my proposed route, and in the end my go/no-go decision was always subject to his approval.

After receiving my private pilot certificate, I learned to make those decisions on my own. Sometimes I learned the hard way. As soon as the ink dried on my certificate I hopped in a trusty little Cessna and flew 600 miles to western Pennsylvania. I never gave a moment's thought to climbing over the Appalachians or dealing with the fast-moving snow squall that caught up with me over West Virginia. Once I tried a high-density-altitude takeoff from a sod field with full tanks, my wife and my dog. I can still hear the stall horn blaring as we cleared the power lines.

Fortunately, none of those lessons resulted in a dime's worth of damage. The greatest lesson I learned was that my private certificate was not proof that I had what it takes to be a good, safe pilot. It was simply a license to learn.

Eventually I got the idea that an instrument rating would change things. Instead, I found myself faced with even more decisions to make and more lessons to learn.

Each time things don't go exactly as planned, I carefully re-examine my decision-making process for that flight. Was there something I could have found out before takeoff that may have changed my mind? Did I miss something in a forecast, aircraft-performance table or evaluation of my own abilities? Did I familiarize myself with all the available information about that flight? Did I make the best choice?

A former commuter captain once told me that there are few absolutes when it comes to making a go/no-go decision. The greatest responsibility any pilot bears is determining whether the benefits of the flight are worth that day's risks, or whether it might be wiser to stay home.

The examiner finally glanced up from his note-taking.

"I'm going to go have a look," I said, tightening my grip on the yoke and sitting up in the seat.

The examiner nodded and went back to his notes. I announced my intentions to the controller. In another five miles I could clearly see that the storm was a monster, and we made a 180-degree turn to our alternate.

Back at home base three hours later, we discovered that the level-five storm had even spawned a few funnel clouds.

The examiner typed out my new temporary instrument ticket.

"You did the right thing," he said. "With 10,000 hours' experience, I wouldn't have done anything different." With that he stood up, shook my hand and presented me with yet another license to learn.

✈ ✈ ✈

*"The greatest responsibility any pilot bears is determining whether the benefits of the flight are worth that day's risks." That's not always an easy decision. The balance of benefits and risks can be a real dilemma. In the following two accounts both pilots were faced with very difficult decisions. Happily, in their cases, they did know better.*

# Better Late Than Never
## by Dominick McCutcheon

Go/no-go decisions—I'd bet you a dollar that pilots who are alive today have so far made the right decisions. I almost made the wrong one.

In April 1986 I was 25 years old and a brand-new VFR Part 135 pilot based in Phoenix, Arizona, flying a Cessna 210 between Phoenix and various Indian reservation hospitals throughout the state. One day I was scheduled to fly with a nurse to remote Tuba City on the Navajo reservation north of Flagstaff to pick up a critically ill infant and return to Phoenix.

Predicting a light load, I topped off the tanks. At the time of departure, there were only three hours of daylight remaining—a serious matter because Tuba City had a high pucker-factor runway: It was poorly lit, unmarked and was surrounded by buildings, trees and power lines.

Departing the Phoenix ARSA, I tuned my com to the reservation pilots' discrete frequency, a kind of 976-CHAT line. Weather was clear, although a squall line was advancing from the west. But if all went as planned, I estimated we would be homebound before the weather deteriorated.

Approaching Tuba City, the surface winds were extremely strong and stirred the dusty painted desert, reminding me of a sci-fi flick about the sandstorms of Mars. I could barely see the ground when we finally arrived over the runway, a beat-up example of what a runway should *not* look like.

On my first approach to land, the crosswind drift and turbulence were greater than I'd expected, so I decided to go around. The next time, with 10 degrees of flaps and a touchdown speed of about 70 to 80 knots, I succeeded in landing, although the navcom radios literally vibrated out of their racks and tumbled to my feet, rendering them useless.

I taxied to the "ramp" (a slab resembling an aborted pouring of a concrete driveway), where a waiting doctor informed me that there would be a delay. A delay would only make things worse, but I explained the deteriorating weather conditions and navcom problem to deaf ears. The doctor gave me a shrug and a "yeah, sure."

Radios are required for 135 operations, not to mention a necessity in getting us back to Phoenix, but I was convinced that I could get by if I declared this flight a medical emergency.

The nurse and I waited, with growing concern, at the airplane. Darkness—and the ceiling—were falling when another doctor and a man with an eerie smile came toward us. When the doctor asked if I could carry another passenger, I said, "Sure, no problem." The doctor pulled us aside and slipped an envelope marked "Confidential" into my hand. The letter in it explained that our new passenger was mentally unstable and had suicidal tendencies.

Bomb delivered, the doctor quickly turned toward his hospital base. This guy could be a big problem in an airplane. Stunned, the nurse and I discussed the best possible seating for everyone's safety: The man should sit copilot with the nurse behind him. That way, if he performed some irrational act, she could grab his head from behind. With him sitting next to me, we could both watch him. It was our opinion that this arrangement was better than having him sit behind me, where he'd be able to grab *my* head.

Where *was* the baby? It was growing darker now and was very cold, windy and snowing lightly beneath overcast skies. I was very nervous about taking off under these conditions, but I rationalized that the 210 was a good airplane. To me, it was the old Chevy Nova of the skies—dependable, simple and strong.

Finally, the screaming baby arrived, oxygen tubes in his nose and an IV in his little arm. The nurse was certainly going to have her hands full on this flight. If it became necessary, there certainly would not be much opportunity for her to restrain our adult passenger. Then, incredibly, the doctor asked me to wait for yet another patient.

Again, I tried to explain the problems we faced relating to weather, radios, oncoming nightfall and overloading, not to mention my unfamiliarity with this part of Arizona. But I lacked conviction and was not firm in exercising the authority of a pilot in command. He left us for a moment, then brought back a Navajo woman who was eight months pregnant. Her luggage and two 50-pound lead X-ray plates also came along. Obviously I was having trouble thinking clearly. Despite the odds against us, I loaded passengers and cargo, motivated by not wanting to abort a mission.

Before I had a chance to yell, "Clear prop!" there was a banging on the copilot's window. Unbelievably, someone was trying to get in. I reached across the man in the copilot's seat and in climbed a 250-pound Navajo man, claiming to be the husband of the pregnant passenger. I looked at my passengers in the ghostly cabin light as the plane rocked rhythmically with the wind. The only sounds were the howling of the wind and the piercing cries of the baby. . . an eerie *Twilight Zone* effect pervaded the cabin. To make matters worse, it was now snowing harder. My passengers gazed at me in complete confidence. I needed to do *something*.

I had not been taught about handling such a situation. A firm decision to go or not to go could only be made by the pilot in command—me—after honest evaluation of the departure and en route conditions: low ceilings, snow, overloaded aircraft, marginal runway conditions, 5,000-foot pressure altitude, and night flight into unfamiliar territory with no navcoms. It may sound trite, but I finally remembered the adage, "If in doubt, wait it out."

Summoning enough courage, I announced dryly, "We are not going." Then I got out. The farther away I walked from the airplane, the more certain I was of having made the correct decision. It was easier now for me to be firm with the doctor who had arranged the flight. There was to be no argument, not even in a case of a medical emergency.

Twice more, the doctor tried to convince me to fly. I was angry and remember making reference to a pilot who, a few years before, had been pressured to fly out of the same airport under similar conditions. That pilot's "go" decision was the wrong decision.

My hesitance about taking control and exercising full pilot-in-command authority could have cost the lives of six people and one unborn. I should never have allowed conditions to grow to such proportions and still have been wondering whether it was possible to fly. Luckily, albeit late, I made the correct decision, the "no-go" decision. I've made subsequent decisions more wisely and far more professionally.

If you are uneasy about *any* conditions of a given flight, weigh them carefully. Only you can decide when your personal limitations are exceeded. Do not allow pride, peer pressure or machismo to make you a statistic.

✈ ✈ ✈

*Young, recently hired, the pilot of the Cessna 210 was very nearly overwhelmed by the pressure of a sick patient and the authority figure of an older man who was also a doctor. It was a decision worthy of a professional, and making the hard go/no-go decisions is how truly professional pilots earn their pay.*

*All kinds of pressures are brought to bear on pilots. A charter pilot has an opportunity to earn some extra revenue if only he can force himself to ignore a sick mag. It's a balancing act that could fall either way. He, too, knew better, and lived to fly another day.*

# Mag Drop Decision
### by John Pratt

The word in my mind as I left the house was WOXO. Indefinite ceiling: zero, sky obscured; visibility: zero. It was difficult finding the road, and the moisture streamed off the car windshield in little swirls and beads. Wipers on, little change in visibility. "A really fine dungeon of island fog," I thought; little need to get to the airport too early. It was July 15; in this Northeastern island resort (Martha's Vineyard), we had a precious few months to make enough to carry us through the long, cold winters, and every day of flying was important.

The rotating beacon was barely discernible as I parked at the terminal and walked into the lobby. The passenger service folks from the regional carrier that served the island were in, but a glance at their board showed everything canceled through mid-morning. Zero-zero it was, and looked like it would stay that way for a while. I didn't bother to go to the hangar and opted for coffee with the guys instead.

At 0930 or so there was a slight brightening of the outside light and some nearby landmarks started to appear. Also, passengers whom the carrier had been unable to reach with news of the cancellations started to arrive, and soon there were five of them milling about and asking how they were going to get to New York. Visibility was inching up, and although the ceiling wasn't much better than 150 feet in fog, we had our takeoff minimums. A call to flight service reported LGA at 800 broken and two miles in fog and haze; not bad and improving rapidly. I filed, signed up my five passengers and headed for the hangar to pick up our Aztec.

Taxiing up to the terminal I paused to do my preflight run-up. Passengers were used to jets, and mag checks and the like can be disconcerting to those who don't care much for "little airplanes" to begin with. I tried to do these empty, whenever possible.

All went well until I tried the left engine mags. A 400-rpm drop and considerable roughness as I switched. This had been a recurring problem with that airplane and was on the maintenance sheet for the next inspection. Up until then it had been cured by leaning the mixture until the engine smoothed out. That day this didn't work. "It's only a mag," a little voice said, "and this is a trip that we didn't count on—an extra dividend, so to speak." Again I tried the switch and again the drop and the shakes. "That's why they have two mags, dummy," the little voice persisted, "get it looked at when you come back at noon."

I was in the act of putting the mag out of my mind and continuing to taxi toward my gate when I looked outside the cockpit. A bare mile in fog and an indefinite 100 to 200 obscured. "You'll be on top at 1,200 feet, LGA is okay and getting better, there's a good, nearby mainland alternate, and by the time you get back the fog will have burned off the island; besides, we need the revenue." He was a persistent little bugger and he had my attention.

With this, another little guy in a white hat woke up and in no uncertain terms said, "That is not the way we do things. The mag is sick, the only hydraulic pump is on that engine, the ceiling is about 100 feet and you will be at maximum gross takeoff weight." I turned around and taxied back to a parking spot outside the hangar.

After driving back to the ticket counter and refunding the five shares of the LGA trip, I returned to the aircraft and with my mechanic and another pilot removed the cowling from the left engine. I climbed aboard and fired it up while they watched. I slowly advanced the throttle and prepared to hit the mag switch. At this moment I saw looks of utter disbelief spread over both faces and a quick "cut" signal across the throat. The engine had failed before their eyes. What connection there was between the mags and the engine failure, if any, I never found out. I did find out enough, however, to ban that one little voice from my head forever.

Single pilot, 100-foot ceiling in fog, six souls aboard, gross weight, and engine failure on rotation, the necessity to retract gear and flaps by hand pump, trying to establish climb in IMC very close to the ground. I didn't tell those five folks that I had given them back more than their money.

That little guy lurks in all of us and if you listen to him, just once, he'll nail your butt for good!

✈ ✈ ✈

*It's hard to know which is more difficult, canceling a trip before takeoff or landing short of a planned destination. The pilot of the Aztec chose not to launch. In the next account, two pilots in a Piper Navajo continued to press on even after they had an indication that something was wrong. When they finally asked for help, it was almost too late.*

# Acid Trip

### by Del Hayes

My colleague Ken and I were on the last leg of a flight, testing electrical equipment in our company's Piper Navajo. We had landed at Memphis to refuel and were on our way again, headed home to Texas. As we leveled off at cruise, I returned to the routine of testing the electronic systems in the aft cabin. It was not long before I sensed a faint, but remotely familiar, odor. As an electrical engineer, I have smelled my share of sizzled systems, and this odor set off my "frying electronics" alarm. But a quick check turned up nothing amiss, and Ken hadn't smelled anything up front.

Just when I began to believe that my nose had been playing tricks with me, I got a stronger whiff. This time Ken thought that he smelled it, too. There were still no other indications that anything was wrong, but to be on the safe side I decided to shut down our test systems. Then I moved up to the right seat.

The situation was more puzzling than troublesome. We agreed that we had smelled something, but there was no smoke, no indication of equipment malfunction and no abnormal indication on the panel instruments. There appeared to be no reason not to press on. But we were in IFR conditions and the possibility of a communications failure, the loss of electrical systems or even a fire made us cautious. Ken told ATC that we might have smoke on board and were going to land at Little Rock, Arkansas.

I had just begun to review the approach plates when it hit us again, this time too strong and pungent to leave any doubt. Something was wrong. We rechecked load meters, shut down avionics and otherwise attempted to determine what was going on. But the origin of the smell remained a mystery. To complicate matters, we began to realize as the fumes got stronger that they were also toxic, and that breathing was definitely becoming unpleasant.

We donned the oxygen masks and refreshed our lungs. Our comfort was short-lived, however, for soon Ken's eyes began to water. It was obvious that the problem was getting out of hand and that we might not make it to Little Rock. Ken called ATC.

"Little Rock, Zero-One Foxtrot is declaring an emergency and would like vectors to the nearest airport. We're descending now."

Little Rock Approach responded quickly with an approximate heading to a small, nearby airport. Ken circled for a short final to the nearest runway, and made a smooth landing despite his teary vision. We quickly taxied to the ramp and were greeted by the local fire department and sheriff. Approach had alerted them that an aircraft was making an emergency landing and might need assistance. Ken cut the engines and we scrambled out of the airplane for some fresh air.

After assuring the local troops that we were okay, we began to search for the source of our trouble. Nothing was obvious, except that the odor seemed to come from the nose. In the Navajo, as in most twins, the avionics boxes are mounted in the nose compartment, so I stuck my head in to have a look. The choking gas drove me back after one short breath.

The problem was the battery, which is also mounted in the nose. An internal fault had caused it to overheat, generating the toxic fumes. If you have a swimming pool and you've ever treated it with hydrochloric acid, you'll recognize the stinging, choking fumes generated by such acids. Of course, the battery is in a sealed case, with vent lines designed to carry any gas overboard, but our vent line was disconnected and the fumes were carried directly into the cabin.

Although pilots of aircraft with NiCad batteries are aware that overheating can occur, most pilots generally assume that lead-acid batteries are benign. This is a dangerous myth. Don't take your battery or its vent system for granted. That goes for singles as well as twins, since the battery of many singles is in the nose as well. Even if your battery is in the tail, air circulation can swirl the gases forward. Inspect the battery box and vent lines periodically. Make sure that the vent lines haven't been clogged, and that the hoses are not loose, cracked or split.

As to the Big Brother aspects of declaring an emergency, a simple report to the local FSDO the next day took care of all the legalities. So pay attention when your sixth sense—or any of the other five—tells you there is a problem.

✈ ✈ ✈

*When they finally got around to it, the pilots of the Navajo were quick to declare an emergency and get their plane on the ground. The next two confessions are by pilots who intentionally went ahead with a maneuver without thinking it through. In the first, the pilot of a McDonnell Douglas F-4, anxious to earn a "Mach 2" pin, exceeded his personal limits—and those of his engines.*

# Streak Phantom
### by Ronald Smeltzer

All pilots have benchmarks in their careers, and for a Navy fighter pilot, hitting Mach 2 is one of them—not earthshaking, but, like the first night loop, it is regarded as an accomplishment. McDonnell Douglas, builder of the F-4, even had its military technical representatives distribute "Mach 2" pins for fast-flying jocks to wear on their lapels.

Not many guys I knew had Mach 2 pins. Going twice the speed of sound wasn't something that had much practical application in a typical military fighter mission. We saw Mach 1 on probably only one hop in 10.

The chances of flying Mach 2 began to diminish rapidly for the pilots in my squadron when Navy directives had us turning in our F-4Js for a new, more maneuverable model of the F-4: the F-4S. The "S" designation stood for *slat*, and that's what made this Phantom more maneuverable. Instead of the F-4J's leading-edge wing flaps, the F-4S had slats along the leading edge. The slats popped in and out as a function of angle of attack; so when the aircraft got slow, the Old Lead Sled could make a pretty sharp turn.

Even when they were in, the slats dirtied the clean wing of the F-4, making for lots of parasite drag. The F-4S also weighed 2,000 pounds more than the F-4J. Because of all the parasite drag, plus the extra ton to lug around, the new airplane didn't have a snowball's chance of reaching Mach 2. Although the airplane could still easily go supersonic, and then some, an F-4S pilot would never win a Mach 2 pin.

That thought crossed my mind as the months passed and we gradually flew the old F-4Js off to either Naval Air Rework Facilities (to reappear later as F-4Ss), Naval Reserve squadrons or Marine Corps fighter squadrons. Finally my squadron had a flight line of shiny new F-4Ss and one lonely, soon-to-be-gone F-4J. I decided I wanted my Mach 2 pin, and if I should happen to get an "unstructured" hop assigned in the remaining -J, I was going to get it.

Shortly before the squadron was due to transfer that old F-4J, I got my chance. I was assigned to fly the jet on a post-maintenance check flight. No hop is more unstructured than a PMCF—

you get airborne, run a prescribed check on the recently repaired system and then use the remaining fuel however you
want. As I recall, we had to check an attitude indicator. My
radar intercept officer for the PMCF was known as Mad Al, who
was about equal in seniority to me and sharp as a tack, a real
go-getter. I knew he was as anxious as I was to hit Mach 2 because we had talked about it some.

After seeing the flight schedule, I scurried around and but-
ton-holed Mad Al. "Madness," I said, "we are going to go all out
with that old bird. It's our last chance to hit Mach 2!" We had a
preflight brief, put on our flight gear and took off.

The flight profile for busting Mach 2 was simple. We would
climb subsonic to around 35,000 feet, put it in afterburner
and unload the Gs to zero. We would go supersonic as we
descended in the acceleration dive. Then we would climb,
this time at Mach 1.2, to around 40,000 feet. Once again we
would unload to zero G and use afterburner to run out to
Mach 2 in the dive. A high-Mach profile was run like that because you couldn't get much above Mach 1.4 in a single acceleration run.

The PMCF portion of the flight went perfunctorily and the
speed run was going fine, around Mach 1.8, when we realized
our acceleration rate was too slow for the airspace we had allowed. We didn't have enough fuel to run around and try again,
and I could see our Mach 2 pins fading fast. In a "one last try"
attempt to hit the magic number, I kept the zero-G pushover
going past my self-imposed limit of a 30-degree dive to around
45 degrees. I figured on pulling out of the dive at 12,000 feet to
avoid flying into the drink.

Suddenly, KABOOM! KABOOM! The airplane shook violently as each engine compressor stalled, flames shooting out
the air intakes. It spooked the heck out of me, because at
nearly twice the speed of sound we had no hope of ejecting
if the engines began to shed parts, and it takes a while to
slow down from such speed. You can't just yank the power
levers out of afterburner, either; that's asking for more trouble.

With the engines chugging and shuddering away, I eased
back the throttles and started the nose up. We leveled off
around 18,000 feet and things settled down. Mad Al was pretty
laid back about emergencies. I'd been wingman for him and
another pilot once when the wing spar in their Phantom
cracked and the last eight feet of one wing just folded up. Mad-

ness simply told his pilot to relax and fly home. However, the mysterious nature of these violent engine stalls had him on edge. Me too, because I had no idea what was wrong with our engines.

Fortunately, we had no further difficulty once we slowed down. After we landed, we quietly told the maintenance officer what had happened. We were concerned that the engines might come apart on the next flight. We all figured they were out of rig, because the F-4's engine hardly ever hiccupped; those two had belched. I drove home without my Mach 2 pin, but glad to be still alive.

The next evening I saw Mad Al at a squadron Halloween party. He and the MO pulled me off to a dark corner and said they had put their heads together about the engine stalls. I hadn't thought anything more of the incident until then, but Mad Al was a meticulous type. They had dug into the tech data on the F4 engine and discovered that at the lower altitude we reached in the 45-degree dive, we were going too fast at Mach 1.8 for the engines to handle the ram pressure of the denser air. If we had kept the dive at 30 degrees our altitude would probably have remained high enough for the engines to handle the load. There wasn't anything wrong with the engines. I had simply flown beyond their capabilities.

It was embarrassing, but nobody else found out. I learned a big lesson, and I think it applies to all pilots—even if you don't fly an F-4.

The lesson wasn't one of engine dynamics or preflight planning of performance factors. No. Our problem arose when I decided, "Just one last try, we're almost there." Such a last-ditch mind-set can bedevil any flying. Accidents happen when a pilot decides to fly just one more leg in marginal weather since he is almost home, or presses an approach below minimums since he is "almost there," or adds one last suitcase. It's always good to grow as a pilot, but be careful the next time you encounter a situation that begs for "Just one last . . . ."

*The Phantom pilot hadn't considered the effect the denser air would have on his engines. The pilot in the next story, ex-FAA Administrator Donald Engen, neglected to plan on the "follow through" as he impulsively flew a Culver Cadet inverted close to the ground—without an inverted fuel system.*

# Administrator's Admission

### by Adm. Donald Engen

For pilots there is often a fine line between being around to tell a story and not being around. This is not to say that aviation isn't safe; it is. But aviation can be terribly unforgiving when foolish mistakes are made or when faulty judgment is used.

I have one of those stories to tell. It used to embarrass me, but now I tell it to show what faulty judgment can do to a pilot. It is a story about a foolish stunt I pulled about three years after I started to fly. That I am here some 41 years later depended on my being able to react properly to the situation, once I had gotten myself in it, and a lot of luck. Like many such stories, it should never have happened.

I was flying a Culver Cadet over the Mojave Desert in California. The Cadet was a peppy little airplane built of plywood and was a delight to fly. This particular one had been modified to a single seat and was to be an antiaircraft target drone. The day was crystal clear, the winds were calm and I was filled with the exhilaration of a good flight. As I approached the airport at Mojave to land, no one was in the pattern. I focused my eyes on the tower to the right of the parking ramp near one of two hangars. I decided instantly to fly by the tower and dropped down on the deck at about 80 feet to pass it on an easterly heading. I had been doing some aerobatics in the Cadet and on an impulse I rolled the airplane as I approached the tower, put in some forward stick and flew by inverted on a level pass.

That would have been fine if the Culver Cadet had had an inverted fuel system—which it did not. The engine quit just as I passed the tower and, as I looked ahead, I still had two of those old half-round, wooden hangars to fly over—deadstick. I passed over the first, inverted, with the airspeed bleeding off. With some very judicious forward stick, I made it over the second hangar by inches.

Now clear of all obstructions, I snapped the little airplane upright and hit the wobble pump. As I was face-to-face with a three-strand barbed-wire fence and about to experience how plywood reacts to prolonged contact with the ground, the engine roared to life. I executed a high chandelle, grateful for the pull of that little propeller, and entered a downwind leg for landing. I never, ever, did that again.

I learned two lessons about flying from that experience. First, pilots should follow the regulations. Second, always think out a maneuver before you do it. Don't perform any maneuver on the spur of the moment. Always plan, whether for a routine cross-country, your first aerobatic maneuver or a night landing.

The adage that aviation is inherently safe but terribly unforgiving is true. I have made my life's work in the sky, and have found the reward of a good flight to be satisfaction and professional pride.

# 8

# Working Together to Manage the Resources

Over the last several years, the concept of CRM, or "cockpit resource management," has become a buzz word in teaching circles, particularly with multiperson crews. Essentially, the theory suggests that if you have someone or something in the cockpit, make the best use of their talents and capabilities.

Captains are taught to seek help and advice from the junior members of their crews. The captain still makes the decision, but he has the advantage of at least considering the input from others. It's a difficult transition for some flight-deck dictators, but the concept has proven its value in actual emergencies.

In some of the following incidents cockpit resource management works exactly as intended. In others, the lack of CRM very nearly leads to disaster. In two of the accounts, the definition of cockpit resource is stretched a bit—but the theory still obtains, use what and whom you have to help in an emergency.

## Vector Analysis
### by L.V. Brandborst

Our G-I was at FL 190, eastbound out of Kansas City for Pal-Waukee Airport, a few miles north of Chicago O'Hare. The warm sector ahead of a weak cold front was circulating Lake Michigan's moisture over northeast Illinois, producing IFR conditions with tops at 7,000 feet. No real problem, although the Pal-Waukee weather was close to localizer minimums.

Although this was 14 years ago, I still remember the descent clearance was to cross the Joliet Vortac at 6,000 and to expect radar vectors to Pal-Waukee. Fine. My copilot ran the approach checklist, hauled out the approach plate and stuck it under my yoke clip. We planned a 120-knot approach for ease in calculating the minimum-descent altitude and missed-approach point, then crossed Joliet at 6,000 feet expecting, as usual, to maintain that altitude across O'Hare's parallel approaches for Runways 14L and 14R.

Instead, we were given a heading of 020 degrees (no problem) and a descent to 2,500. The copilot questioned me with a puzzled look. After all, we were still some 45 miles from Pal-Waukee. I briefly considered asking to maintain 6,000 but felt that there was a good possibility of being vectored halfway back to Kansas City. Fuel was no problem, and our 200 knots indicated would eat up those miles in a hurry. I nodded and we sank deeper into the murk.

Though we were not monitoring O'Hare's other approach frequencies or its ATIS or navaids, it seemed obvious to us that the ILS approaches to Runways 14L and 14R would be in use. We had blind faith in our controller, though we seemed to be the only traffic he had on that frequency. Staring straight ahead into the void, I began my rehearsal of the Pal-Waukee headings, altitudes and times.

Then, in the lower-left corner of my windshield, the floodlit vertical fin of a TWA 707 appeared. It passed left-to-right, under our nose, ghostly quiet, and vanished. My copilot turned to me.

"You see that?" I asked. He nodded wordlessly.

"Confirm our altitude and heading," I said, finding it difficult to remain calm. Approach control confirmed our altitude and heading.

"Uh, Chicago, we just passed over the tail of a southeast-bound TWA 707."

The silence on the frequency seemed interminable, then, "Roger, turn left heading 330, vectors to the final approach course at Pal-Waukee," was the reply.

It was easier to maintain a certain aplomb after the urgency of the situation had passed. After landing, and just before shutting down to unload, ground control asked me to come to the office for a phone call. "Captain, this is Smith [I have forgotten his real name], Chicago Approach. I have the TWA captain on the line."

A youngish-sounding confident voice: "This is Captain Jones [I have forgotten his name, too]. It was our fault." He said *our* not *my*. "We were assigned 3,000 to the marker and just started descending at glideslope intercept." A pause. "We didn't see you."

The controller came on, "Captain, the ball is in your court." My copilot had come into the office and was standing by my elbow, eyes questioning.

Many thoughts crowded my mind, not the least of which concerned my job security. If I were to report a near-collision on the approach path to the world's busiest airport, surely the *Chicago Tribune* and TV stations would raise a hue and cry for better and tighter air traffic control. I was 52, nearing the end of my career, and hoping to close it out in a new G-II. The TWA crew and the controller might have had their unions handle their litigations, but the FAA would have issued at least suspensions. We could, as a group, be held morally responsible for passengers who might have second thoughts about flying on the airlines or on corporate airplanes.

The controller persisted. "Captain, what do you want to do?"

What if I had been two knots faster and 10 feet lower? It would have been me, my copilot, the CEO of a 30,000-employee major corporation, several division presidents and perhaps 200 TWA passengers and crew spread all over Arlington racetrack.

"Oh, let's forget the whole thing," I finally said. "Are you sure?" someone asked.

I repeated my decision, said goodbye and hung up. My copilot stared at me with wide-eyed disbelief, as if I had refused a winning lottery ticket on religious grounds. To his credit, he said nothing. Later in our hangar in Milwaukee, as I sat at my desk trying to ignore a stack of Jepp revisions, the copilot came in with two glasses filled with amber liquid. He knew I treasured a good Manhattan, but he was a teetotaler.

"I didn't know you drank," I said.

"I do now," he answered with a tentative sip. "Why didn't you ask for a lifetime pass on TWA," he asked, "or at least a trip to Paris?"

In retrospect, three people made three major mistakes. The controller's was in releasing us to minimum altitude too far out, to cross final approach paths near a glideslope altitude. The TWA captain anticipated the usual descent on glideslope, wrongly leaving an assigned altitude at glideslope intercept. My own mistake was accepting the clearance in the first place and not challenging the controller in my haste to complete the flight by avoiding time-consuming vectors.

I hope that somewhere in the TWA system there is a crew still active, possibly still disbelieving its good fortune at the hands of a gullible corporate crew. I hope that those men also learned a lesson in concentration. I would like to think they are not the same ones who hit a ridge north of Dulles International Airport a few years ago after leaving an MEA too soon. I also hope that the controller survived Reagan and PATCO. For my part, I ran out my career (the company did get a G-II), and I have come full circle in aviation, now running a ground school in Arizona.

My students seem to absorb this story as an object lesson, but older, more experienced pilots often dismiss it as something that happens daily. I am certain that many of them refuse to report their near-collisions for reasons similar to mine. I feel better for having kept this secret officially intact over 14 years, yet relieved in telling it publicly for the first time. It may be the reason why, on subsequent approaches, I hitched the seat a bit forward, scanned the sky with zeal, and insisted the copilot do the same, even if he asked, "Why bother? You can't see anything anyway."

✈✈✈

*"The copilot questioned me with a puzzled look." At that point the pilot should have considered his and the copilot's hesitation about the clearance they had been given. The breakdown of CRM in the cockpit of the airliner on the approach also contributed to what could have been a disastrous event.*

*When CRM works, it can get pilots and passengers out of a situation that might otherwise be unsurvivable. Taking off in a grossed-out Boeing 747 "is an exercise of faith" that everything will work the way it's supposed to. If it doesn't, then it takes CRM and a miracle. . . .*

# Stalling for Time
## by Julian Kulski

The air was unstable; we knew that from the gusts on the precarious climb up the jetway. There were scattered rainshowers in the area, with the prevailing visibility more than 25 miles.

Gatwick's ATIS had the temperature at 40° F, with winds from the southwest at 20 knots and an hour-old report of wind shear from an aircraft on final for Runway 26L. During the taxi to 26L, dispatch informed us that our takeoff weight, including freight and fuel to fly to Miami with reserves, was a little more than 730,000 pounds; we were 1,000 pounds under max gross weight for our runway. We had 442 souls strapped down in the back. At 10:57 the tower gave us winds from the southwest at 10 knots and cleared us to go. Getting a grossed-out Boeing 747 airborne is an exercise in faith. Although the book says you can lose an engine after V1 and still make it over the hills, a prudent soul would prefer not to try. At 151 knots the first officer called V1, then VR at 160 knots. Our next goal became V2, 166 knots.

The moment of rotation is a time of watching and waiting: watching the gauges; waiting for the airplane to decide to fly. At the same time that our main gear cleared the runway we heard a loud bang from the aft and starboard. My first thought was that we'd hit something. But immediately the number-four engine EGT went off scale, with its amber warning light a pathetically inadequate signal of an enormous problem—compressor stall. The stick-shaker went off.

As I reached for the power levers, both the captain and the copilot called for power. I continued right to the stops on engines one, two and three, which all howled in agreement. Because of the captain's flying skill and the three working engines, we flew—but just barely. We flew on stick-shaker for about 30 seconds with the airspeed fluctuating in turbulence between V1 and VR. Two miles off the end of 26L, Russ Hill rises more than 200 feet above the airport. Though we seemed to observers to hang in the air, our groundspeed was covering 250 feet per second toward the hill. As the copilot raised the gear I flipped open the jettison panel and palmed the toggles to dump fuel out the wingtips.

The captain was alone now. Our radar altitude showed us at 260 feet agl 20 seconds after rotation, about a mile from the runway end. Twenty-nine seconds later, or about two-and-a-half miles from the end of the runway, our radar altitude was 113 feet. At our deck angle the tail was barely clearing the treetops. We were hanging on the fans in ground effect, fuel pouring out both wingtip dump tubes, smoke and flames trailing from number four and, we later found, number one.

Some 70 percent of the engines' thrust comes from the massive compressor fans out front, and they are susceptible to compressor stall in strong crosswinds. While number four had fully stalled, apparently number one was thinking about following suit. Though we had no indication of it in the cockpit, witnesses on the ground and in the airplane later said they saw smoke and flames from both outboard engines. It took a while, but once we'd cleared Russ Hill the captain nursed the pitch along and achieved V2. In England ATC maintains radio silence during an emergency; the first officer now broke this silence to tell ATC that we'd need vectors for 35 minutes of fuel dumping before we could return to land. Seventeen miles from the end of the runway we were at 1,500 feet agl.

After dumping 160,000 pounds of fuel we returned for a normal, albeit somewhat tense landing, and at 12:04 we blocked into the gate. We knew we'd been low, but just how low was first made clear by the ashen-faced tower controllers who suddenly loomed in the cockpit. They'd seen us stagger over and disappear behind Russ Hill with trees waving in our wake and called in the crash alarm, only to see us emerge as a welcome blip on their radar screen. A passenger later said he'd looked through the parted curtains in the second-story window of a Georgian mansion.

Hours later we debriefed with an investigator, who would ultimately focus the blame on inadequate wind sensing at Gatwick. What I realized from the incident was that the crew concept had worked. Prior to that day, had a simulator instructor programmed an engine failure on rotation at max gross weight along with wind shear, I would've cried foul. But Lawgiver Murphy doesn't care about fouls.

✈ ✈ ✈

*The crew concept worked. Each crewmember had contributed and performed as required and directed by the emergency at hand. Not all captains have gotten—or accepted—the word about using their crews. In the following confession a rookie on the line learns a valuable—but unintended—lesson from a senior captain who insists on doing it all himself.*

# Saved by the Wipers
### by Maurice R. Azurdia

As a new hire flying DC-9s for a domestic airline, I was in seventh heaven, ever eager to please my captains. With only one month of experience flying the DC-9, I drew a trip with a captain who, despite the fact that he was number one on the line's reserve list, reported in cursing them for calling him out to fly. He also complained continuously about the working conditions at our company, conditions which I found more than adequate, and was generally in a bad mood during the entire trip.

A rookie on the line, I was anxious to learn as much as I could from my assigned captains, and the majority of them tried their best to teach me the peculiarities of the DC-9. Not so this individual. He totally ignored me and exhibited a disregard for company procedures.

On the final leg of the trip, the leg I would be flying, we were approaching a major airport in terrible frontal weather. It was a hot, humid summer night and the destination ATIS was reporting rainshowers and thunderstorms. We ran through our checklists and were vectored by ATC for an ILS approach to Runway 30. At our initial approach altitude of 3,500 feet msl, we were in intermittent rainshowers with continuous light to moderate turbulence, and our decrepit radar system was becoming quite useless due to attenuation.

All the while, I felt confident because I was working with a professional. Or so I thought. Then ATC called and, in a friendly voice, advised us of its intentions to keep us high just a "tad longer" because we had a light VFR airplane crossing ahead and below. I wondered who was brave enough to venture out in such weather in a light airplane, then became increasingly concerned because we were now intercepting the glideslope from above. I knew I should start down, but the controller had us holding altitude.

I slowed the fully loaded DC-9 as much as possible. The controller had not yet called us back with any altitude changes, so I asked the captain to remind ATC of our altitude restriction. The controller was honestly surprised to get the radio call; he had forgotten about us. Quickly, he asked if we could still "get down from there."

The glideslope needle was pegged at the bottom of the Collins FD-108 flight director. But the captain said, "Yes, we can do it," and reached out and extended the gear—without ever including me in his plans. He pushed forward on the yoke, ordered me to start down, and retarded the throttles to idle while I was still holding them. Then he began feeding in the flaps until we were in landing configuration—in heavy rain and moderate turbulence, dropping to earth like a tractor trailer.

The intensity of the rain increased, so much so that we had to shout at each other to be heard. "I don't like this!" I yelled. He ordered me to continue down to "take a look." If it didn't look good, he said, we'd get out of there. To me, it was already looking as bad as I ever wanted it to get, but being new and inexperienced, I obeyed the voice of The Boss and continued down, at close to 3,000 fpm.

The rain was now very heavy, and we activated the windshield wipers. The ground proximity warning system (GPWS) detected our high sink rate, and we got the *whoop—whoop, terrain! pull up! pull up!* warning. At 500 feet agl everything came apart at once. The glideslope shot up from the bottom of the instrument—way too fast to catch it—and when the approach lighting system became visible through the heavy rain, I knew we had to get the hell out of there. The angle of the approach lights clearly indicated that we were too high and too close to the runway to attempt a landing.

The captain obviously did not agree with me. He, too, saw the approach lights, yet was bent on a landing. But at that precise moment the captain's windshield wiper failed. He was unable to see well enough to land, so he asked me if I had enough of a view to put her down. My reply was to advance the throttles to go-around power, rotate the nose of the airplane and call for flaps 15. I have no doubt he would have attempted the landing himself if his windshield had been clearer.

The wonderful night was not yet over, as I had executed the missed approach directly into a storm cell that was sitting at the end of the runway. For the next five minutes, I fought to keep the DC-9 right-side-up. Although I had been trying to convince the captain to divert to our alternate, he advised me that he wanted to try another approach. I could not believe my ears. The only thing that saved us and our unsuspecting passengers was hearing the calm voice of another captain from our line talking to ATC. Following behind us, he had also missed the approach, and requested clearance to his alternate. My captain listened to the old dog retreating to fly another day, and that—finally—made him agree to divert. Later that night we were able to return to our original destination in smooth air.

My lack of experience, coupled with the attitude of this captain, nearly ended my airline-flying career in a ball of fire. The captain in question recently passed away from natural causes.

And I have since learned to speak up, loud and clear, in the cockpit.

*As the older generation of pilots retires and younger pilots move into the left seats, cockpit resource management is fortunately becoming more prevalent. But there are still situations in which the best trained crews backslide and forget the importance of first flying the airplane. The two-man crew of a DC-9 in the next account got caught up in a passenger's emergency and almost incurred an emergency of their own with their airliner.*

# Heart Burn
### by James Deeton

The cockpit door opened and a flustered flight attendant blurted out, "We have an elderly lady who's having a heart attack!" We were descending into Atlanta on a hot, muggy mid-afternoon in August.

"See if there's a doctor or nurse on board. We'll call ahead and have someone meet the airplane," the captain said.

"There aren't any, we already checked," the attendant said. "I'll see how she's doing and get right back to you."

It was the captain's leg to fly. "Call the company and tell them we have a heart attack victim," he said to me. "Have paramedics and an ambulance meet the airplane. Tell them we should be on the ground in about 15 minutes. I'll see if ATC can get us right in."

Approaching an airport is a busy time for the two-pilot crew of a DC-9. Fuel, pressurization and hydraulic systems must be reconfigured. Radio communications and frequency changes increase dramatically. Checklists must be accomplished and read. The ATIS must be listened to; altimeters are reset descending through Flight Level 180.

The instrument approach is reviewed and orally briefed. Visual surveillance for conflicting traffic increases; and heading, airspeed and altitude changes come fast and furious as approach control vectors the airplane for landing.

I performed all my regular duties, conferred with our flight attendants several times about our passenger's condition, and arranged for medical assistance to meet the airplane. Things were happening fast. Approach control vectored us straight to the airport and deleted the normal 250-knot speed restriction below 10,000 feet.

From our flight attendant's reports, the heart attack was no false alarm. Getting the airplane on the ground, soon, seemed an excellent idea.

Approach control vectored us to the outer marker for an ILS approach to Runway 8. Our radar painted a fairly heavy rainshower just inside the outer marker moving toward the airport. It looked like we and the rainshower would arrive at the airport together.

I was in my eleventh year of airline flying. I'd spent almost 5,000 hours at the flight engineer's panel of a DC-8 and a Boeing 727, and about 1,000 hours as a 727 copilot. I had a little more than 300 hours in the right seat of the DC-9.

We entered the rain within seconds of passing the outer marker. The gear was down with three green lights, the antiskid armed, the seatbelt and no-smoking signs on, the flaps at 40 degrees and the annunciator panel checked. We were ready to land.

"Before-landing checklist complete," I said. The rain became a roar.

"Turn the wipers on high," the captain said.

The rain grew louder, drowning out the engine's whine. The only sounds that rose above the din were the frantic rhythmic thuds—like two metronomes gone mad—of the windshield wipers.

Descending through 1,000 feet the captain called, "Out of a thousand." I looked through the windshield and saw only opaque gray.

At 500 feet the wind shifted. The captain reduced power to stay on the glideslope. Our rate of descent increased. We now had a tailwind. A voice in my head asked, "Go around?"

At 300 feet I began to see the sequence flashers. "Rabbit in sight," I said.

The approach lights materialized. "Approach light straight ahead, runway in sight," I called out.

Our airspeed bugs were set to a VREF speed of 121 knots. The captain had held that, plus the 10 to 15 knots extra the weather conditions dictated, throughout the approach.

We touched down, left main first, on the 10,000-foot runway, 200 feet beyond the 1,000-foot mark. Our airspeed was close to 120 knots, but our groundspeed was probably higher.

The spoilers deployed automatically and the captain pulled the throttles into reverse. It was raining heavily.

"Two lights, four lights," I said, as the engines went into reverse. Two amber lights showed the reverse mechanisms were unlocked, and two blue lights indicated the engines were producing reverse thrust. "Ninety percent on the right, 80 percent on the left," I said. The captain evened the power on both engines to 80 percent.

I pushed the yoke full forward with the wings level.

When we had first touched down the airplane felt as if it had accelerated. A third of the way down the runway it was alarmingly evident the airplane was not decelerating as it should.

The captain braked and reached for the nosewheel steering wheel with his left hand.

We started to skid. A queasy, unattached sliding feeling invaded my stomach. The airplane began to snake along the runway's centerline, the nose oscillating from one side to the other as the captain fought to keep it straight.

Halfway down the runway I heard a muffled pop—a blown tire. Two-thirds down the runway there was another pop—a second tire blew. We had slowed, but the airplane still felt as if it were about to spin out from beneath us.

I looked at the airspeed indicator. "Ninety knots," I called out. It was still raining hard and the end of the runway was getting bigger.

The boom of compressor stalls sounded as the engines ingested their own hot exhaust. Normally, reducing reverse thrust would cure this, but it seemed to be the only thing going for us, so the captain kept the engines in full reverse.

The airplane wasn't going to stop. We would run out of ideas and runway at the same time. With 1,000 feet of runway remaining, emergency evacuation came to mind.

My manual said I was supposed to evacuate through an exit in the forward area, taking the cockpit fire extinguisher with me. I was then to take command in the forward area and assist in passenger evacuation.

Then with less than 100 feet to go, the airplane stopped.

We turned off the runway. It took more power to taxi and the ride was lumpy. We attributed that to the two blown tires. The tower asked if we needed assistance, but we declined. We wanted to get our heart attack victim to the gate.

Waiting for us at the gate were an ambulance with flashing lights, a company station wagon, two paramedics with a stretcher, several people in suits holding clipboards and three gate agents.

The captain stopped the airplane, set the parking brake, shut down both engines and pushed his seat back.

"Where's our lady with the heart attack?" he asked the flight attendant as he opened the cockpit door.

"She's right here," she answered, pointing to a small, gray-haired woman standing in the galley.

"I'm sorry if I caused you any alarm. I guess it was just indigestion," the woman said, patting her chest and smiling sheepishly.

After the confusion settled down and the passengers were gone, I went outside to look at the airplane.

Two mechanics in yellow slickers squatted near the left main gear. They turned and looked at me. One smiled grimly and raised his eyebrow. "It must have been quite a ride," he said.

The left outboard and right inboard tires were blown. An X was carved transversely across the treads. The tires must have exploded. I looked down into the holes in the tires and could see the wheels' axles. Fuse plugs on the other two tires had melted and air hissed out of them. Tendrils of curly, spaghetti-like rubber hung off the flat spots on the tires' surfaces.

We had hydroplaned, a phenomenon that occurs when a tire's tread surface doesn't make firm contact with the runway. I learned more about hydroplaning than I ever wanted to know.

I also learned firsthand that if the approach and landing don't look or feel right, for whatever reason, go around. The onboard medical emergency we thought we had was a compelling and seductive reason to get the airplane on the ground.

But if you balance one heart attack victim against careering off the runway, and the inevitable injuries and damage that would be sustained in any emergency evacuation of an airliner, the right choice becomes clear.

*Cockpit resource management has been expanded to include virtually anything inside the cockpit as well as some outside resources as well. Controllers, mechanics, or instructors on the ground, available by radio, can be valuable resources in a cramped cockpit during an emergency situation.*

*As the remaining incidents in this section illustrate, CRM should not be limited to pilot crewmembers. Anyone, pilot or passenger, can be helpful as a resource to aid the pilot in an emergency. Sometimes they can also cause the problem.*

# Wrecks, Dives, and Videotape
## *by Daniel Sauer*

All was calm that winter morning until the phone rang. It was the chief pilot for the small FBO where I worked part-time. He wanted to know if I would take a reporter and a cameraman for a ride around the airport in our Piper Aztec and shoot an ILS to a landing. The weather had been below minimums for a couple of days earlier in the week, and a light twin had crashed while trying to get in on the same ILS. The weather this morning was just above minimums.

My boss explained that the TV crew wanted to film the actual ILS approach and instrument indications to show the public what goes on during an instrument approach. I was still not convinced—there had been some shoddy aviation reporting done in the past by the local TV stations, but I finally relented when my boss told me that the reporter was highly respected for his factual reportage.

I stopped at the flight service station to make sure I had an alternate close by, in case I couldn't get back in or an emergency came up. Once at the FBO I briefed the reporter and cameraman on what we would do.

After a thorough preflight we boarded the Aztec with the TV crew in the center row of seats and taxied out to the ILS runway in use. I performed the run-up and takeoff checklists and showed my passengers what instruments I would be monitoring to guide us along the localizer and glideslope of the ILS.

I advised the cameraman that during the takeoff and landing he would have to stow the camera, since I was not interested in wearing it as a headpiece should we come to a quick stop.

Without further delay we were cleared for takeoff. Soon after liftoff we entered the gray murk and accepted vectors from departure control for an ILS to the runway we had just departed. We were climbing through 500 feet agl when the camera started rolling.

As we passed the outer marker on downwind at 2,500 feet, I engaged the autopilot so that I could explain more readily what was going on and point to a few of the instruments I was using. As I received my final turn to intercept the localizer I flipped the autopilot switch from heading to localizer and monitored it for the intercept. I was cleared for the approach and was turning to intercept when I turned around for a few seconds to speak with the reporter. When I returned to my scan I was startled to find I had gone through the localizer and was descending rapidly in a 45-degree bank through 2,200 feet, well below the glideslope.

Approach control was just telling me that I was to the right of course centerline when I disengaged the autopilot with the control wheel switch and started a wings-level pullup. The controls felt as if the autopilot were still engaged, and it required great amounts of pressure to override it. I did not get the airplane back under control until I had shot up to 3,000 feet on a heading quite different from the inbound course.

Until that point there had been a running conversation among myself, the cameraman and reporter. Now it was as quiet as a mummy's tomb and you could almost hear the three of us thinking out loud, "My God, we almost became the second airplane in a couple of days to crash on the same ILS, probably within feet of each other." And our accident would have been captured on film for all to see.

I told approach control that I was having control problems and requested a westerly heading to check it out.

The controls still felt much heavier than normal, and I told the passengers that I thought the autopilot had malfunctioned. I double-checked that the autopilot was off by looking at its control panel—the switches were all off.

So I decided to flip the heading switch back on to test the response. As soon as it was switched on, the airplane started to bank steeply. I shut the autopilot down again and watched the switches click off this time. It seemed that the autopilot was still on anyway. There must be a short circuit, I thought, so I reached over to pull the autopilot breaker. Instantly I realized what had almost cost us our necks—the copilot's seat back had been pushed forward by the cameraman so that he could get a better shot of the instrument panel. Normally this would not have been a problem, but the headrest had been removed because it was broken, and this allowed the seat back to become wedged under the control wheel when power was reduced for the approach. I put the seat back upright, informed the TV crew of the problem, tested the autopilot and told approach we were okay and requested another ILS.

The approach went along without a hitch, and we broke out about 50 to 100 feet above ILS minimums and executed a satisfactory landing. When we shut down I asked if there were any questions and if they were going to mention our escapade on the evening news. They said they understood what had happened, and since they were partially to blame for it, they would keep it to themselves.

Closing up the airplane I looked at the top of the leather seat back and could see where the control wheel had dug deep cuts into the leather from the back pressure I had exerted. From that moment on I decided to be more aware of what my passengers are doing while I'm in the left seat.

*For pilots, managing their resources means being aware of what others in the cockpit are doing and being sure they are not hampering the progress of the flight.*

*In the next tale, a Cardinal RG pilot was able to marshal his passengers and experts on the ground in an unusual way to correct a landing gear extension problem.*

# Fancy Footwork
### by Alfred Pinkerton

The preflight check on the Cessna Cardinal RG had been normal, and we rolled for takeoff at approximately 7:30 a.m. on our way to do some water skiing for a few days at nearby Lake Powell. My passengers were a Los Angeles fireman and his fiancée, a math teacher. As we neared rotation speed, the nose of the airplane suddenly dipped, and I instinctively applied back pressure on the yoke. The nose came up, and we lifted off the runway with the stall warning blaring. I let the airplane back down on the runway, and within a second or two we made a normal takeoff. Needless to say, we were all wondering what had happened.

After a quick scan of the gauges, I reached for the gear switch, only to find it already in the up position. It suddenly dawned on me that we might have a real problem. But I wasn't getting a "gear up" light. I cycled the gear to no avail and requested to remain in the pattern for a low pass by the tower. Tower reported that the nose gear appeared retracted, but that the mains were hanging about halfway through the retraction cycle. No doubt about it now. . . we had a real problem.

We departed the pattern to get our thoughts together. I asked the tower to call the co-owner of the airplane to bring the service manual to the tower. Then tower gave us our own frequency. The good part was that we were in no immediate danger. The weather was good, and we had seven or more hours of fuel. We went through all of the normal emergency procedures for lowering the gear. While this provided us with "busy work," it did nothing to help. Next I tried some abrupt maneuvers to shake the gear down and only succeeded in turning my passengers green.

The local FBO, who performed regular maintenance on the airplane, sent one of its shop foremen over to the tower. We now had a professional team on the ground ready to go to work on the problem. With some really basic tools that we had on board, we were ready to begin disassembling the aircraft from the inside out.

Since I was the only pilot on board, one of my passengers, the fire captain, was elected to the position of airborne aircraft mechanic. Piece by piece we took the floor apart until we had all the working parts of the gear system exposed. At that point we were able to determine that the gear box that drives the gear system was completely wrecked. When the airplane had settled back on the runway just before takeoff, the main gear had evidently already started to retract (with the gear switch in the UP position, the gear starts up as soon as the weight is off the nosewheel) and contact with the runway sent all of the weight of the airplane directly into the gear box. At this point there was absolutely no way the main gear could be moved by using the gear-retraction system. We now had to look for some other method to lower the main gear.

Meanwhile, the ground crew suggested we hang outside the airplane and shove the gear down with a good kick. Well. . . this was getting to be more of an adventure than we had bargained for. We asked if this procedure had been successfully done before or if it was just a theory. The ground crew came back with a very casual "We understand it's been done before." I was fortunate in that fire captains don't seem to be too concerned about height. We didn't want to make a bad situation worse, but we didn't want to risk a belly landing if there was another way. Finally we decided to go for it, being very careful not to drop our fire captain into the middle of Disneyland from 3,000 feet.

We slowed to a minimum safe flying speed and trimmed the airplane. As my passenger eased out the door, I hooked my right arm through his knee, and he kept a very firm grip on the base of the right seat. We were an instant success. He brought the gear forward and slammed it home. Before we could finish congratulating ourselves, however, the gear went back where it had been in the first place. We tried putting the gear switch in the DOWN position, and our fearless fire captain went out again. The right gear was down and locked in moments. Then he went for the left main, but it was solidly locked in the trailing position and refused to move. Now we had the nosewheel and right main down and locked, but couldn't move the left main. That wouldn't make for a smooth landing.

The problem is that with the gear switch in the DOWN position the hydraulic ram that once operated the gear tries to do its job. With the ram extended, the gear box jams and the main gear is wedged in its trail position. We decided that the gear box would have to be completely disassembled. Taking the floor up with the tools we had was one thing, but the gear box presented a much bigger challenge. However, by this time we were so far into this project that nothing seemed impossible, although the two hours that had passed since our takeoff were beginning to show. Eventually the gear box gave in, and we could operate the gear switch without jamming the box (or what remained of it).

We cheered on our airborne mechanic and felt confident that this time the gear was down and locked and would stay there. Our human gear-retraction system readied himself for another trip into the wild blue. The right side was down and locked (and double-checked with a good solid kick). The left side gave in and was also double-checked with a kick. I looked down and saw the green light indicating "gear down and locked."

After two fly-bys past the tower, we were told that the gear looked good. About a mile out on final we opened the doors—just in case—and as we crossed over the threshold we shut off the fuel. The landing wasn't the best one I've ever made, but the sound of the wheels rolling along the pavement was real music. The engine stopped as we rolled off the runway onto a taxiway, and we came to a stop three and a half hours after we took off.

How did the gear switch get into the up position for the takeoff roll? I really don't know. Either it was in that position when I did the ground check and I missed it, or it was inadvertently moved when we were loading the airplane with our water skis and other luggage.

Why did the nose dip when the nose gear retracted? I had formed a habit of trimming the airplane slightly nose low, so it would stay on the runway until I wanted to move it off. I don't do that anymore.

There really is a moral to this story. As in many aircraft accidents that I have read about, it is the little things that can jump up and grab you. When you see an aircraft sitting there on the ground, there is no question that the gear is down. It's easy to become complacent in checking every item as thoroughly as we should. After all, we have done this hundreds of times without a problem, haven't we?

*The pilot was lucky when it came to letting it all hang out that he had a fire captain on board who was used to heights. The next example in this chapter is an excellent illustration of how a pilot can enlist resources that are not necessarily on board the aircraft. The incident involves a de Havilland Twin Otter crew that recruited a helicopter to serve as part of an instrument landing system. Talk about managing your resources!*

# The Triumph of Hope
## *by Tom Hammersley*

The morning dawned cool and gray in the mountains of northern Ethiopia. The sun had not yet appeared above the horizon, but the sky was beginning to glow red.

The country was being ravaged by a famine of catastrophic proportions that was caused by a 12-year drought. To complicate matters, a civil war was keeping the government in turmoil. The organization that employed me was responsible for providing logistic air transport for the multifarious relief agencies in the interior. Here at Alem-Ketema we were under contract with the Southern Baptists to shuttle grain across a 30-mile-wide canyon, the Jema River valley. Moving the grain across on mules had proved too slow, and seven-ton Mercedes diesel trucks were no match for the roads. In time it was determined that moving the foodstuffs by air would be the most cost-effective method.

The airplane of choice was the de Havilland Canada Twin Otter—no finer aircraft could be used in the rugged interior of Ethiopia. The Twin Otter could haul 2.2 tonnes per crossing. Its two PT6A-27 engines, allied with its STOL capability, had enabled us to operate in and out of these rugged mountain airstrips with relative ease. We were the lifeline to more than 30,000 people at Guna-Meskel, a village on the opposite side of the Jema River canyon.

With the preflight completed we started the engines, taxied to the airstrip, and embarked on our shuttle flights for the day. Hungry but hopeful people were already gathering on the other side.

By midmorning some scattered rainshowers and low stratus clouds had moved into the area, and poor visibility forced us to consider curtailing the shuttle. At 10 a.m. we were taking on another load of grain, and more fuel, when a gut feeling told me to put on an additional 100 pounds of fuel. A normal fuel load for the round trip was 500 pounds, which provided a 30-minute reserve. The other side of the canyon was still visible through the light rain and mist, and we taxied out and began a flight that will remain a vivid memory for the rest of my life.

My copilot, David, eased the Twin Otter into the air once again, making a gentle left turn on course toward Guna-Meskel. Light rain tapped on the windshield, and wisps of cloud streaked by my window as we slipped into the mist.

Most of my 5,000 hours had been accrued flying as a commuter pilot on the east coast of the U.S. Having flown around the Delaware and Chesapeake Bay areas most of my life, I was well acquainted with the hazards of low visibility due to light rain and fog. I had never seen the weather change as rapidly as it had on this June morning. In the few minutes it had taken us to cross the canyon, a layer of low stratus clouds had moved across the entire area, covering the escarpments and mountain-sides. Above the clouds the sun was bright and the sky blue, interrupted by gray snowcapped mountains in the distance. Directly below was the crocodile-infested Jema River and the rugged valley floor.

David looked over at me and said, "Well, what do you think?"

"Circle the plateau to the right," I told him. This would put the airstrip off to the left of the aircraft, giving me the best opportunity to see it out my side window. The horizontal visibility had become even worse, perhaps three-quarters of a mile at best. After circling, we found no way to make a safe approach to landing.

"Let's take it back and wait out this weather," I said.

But back at Alem-Ketema the weather was no better, and the ground personnel there told us over the radio that a landing would not be possible.

"Let's turn west."

"Going to put it down at Maranya?" asked David.

"That's our alternate strip," I reminded him. "We can have lunch with the nurses at the clinic while we wait out this weather." Maranya was only 12 minutes away, and I believed we would find refuge there. We skirted through a mountain pass, the most direct route to our alternate strip. As we came out the other side of the pass my heart sank. Directly ahead was a large thunderstorm cutting us off from all safety. That's when I took over the controls.

"Is there anything you want me to do?" my copilot asked.

"Yes," I said. "You can begin to pray."

"I already have," he smiled.

I turned the aircraft back toward home base. Attempting a landing in the blind would be risky, but at least there would be immediate medical attention at Alem-Ketema. The fuel low-level lights glowed an ominous red. We were in grave danger.

A quick mental calculation told me that we would be back over AK with about five minutes of fuel, enough to try one approach; I would have to put it on the ground the first time. With no radio navigation aids to guide us down, finding the runway would be little more than guesswork, but it was our only chance.

Instruction from my days as a student pilot suddenly came back. I remembered the four "Cs"—climb, communicate, confess and comply. First you climb for better radio reception, as you attempt to raise someone on the radio who might be of some assistance. When you establish communication you then confess your situation and comply with their instructions to assist you. How fundamental, I thought.

"Emergency boost pumps on," I told David.

"Check two on."

I eased into a shallow climb, not wanting to flame out the engines, and began to make a distress call. It was picked up by Grant Louding, a helicopter pilot slinging grain at Rabel, about 25 miles to the north.

"Grant, our fuel situation is critical," I told him. "We're approximately 10 minutes from AK."

Grant's voice was barely audible, but it crackled through.

"I'm on my way. Estimated time of arrival over AK is eight minutes."

With approximately 10 minutes of fuel left on board we were still five minutes from the airport. Grant's plan was to fly toward us and intercept our inbound course. From there he would be our eyes to guide us down to the runway. The helicopter has the distinct advantage of being able to hover and move across the ground at a slow speed. He could feel for the runway himself, while we could line up on him to give us an idea where the runway would be.

Grant flew the helicopter with great precision. He inched it along the edge of the plateau, groping for the approach end of the runway in the thick fog. We were two minutes behind him, in a slow descent for the approach.

"I'm over the runway," Grant said.

He swung the helicopter around to face us on our inbound course to the runway and turned on every light he had. To us he appeared as a single white light in the mist.

I slowly banked the Twin Otter onto final approach.

"Tighten your seat belt and pull your shoulder harness down snug," I instructed. "I'm going to put it on the first time." There was no fuel for a second chance.

*I will never leave you nor forsake you,* came a still, small voice from within. My mind flashed back to Bible study in my youth as we strained hard, peering through the windshield, hoping for a glimpse of the runway. Suddenly, to our amazement, the fog seemed to roll back from both sides of the runway, only enough to see the asphalt strip. But there it was ahead of us in plain view. I easily set the aircraft down and quickly brought it to a stop. In the short time it took to turn around, taxi in and park the airplane, the entire airstrip had fogged in again.

I felt we had been delivered. I climbed out of the airplane to the laughter and tears of many of my co-workers. David had climbed into the Jeep for the ride back to camp.

"Come on, Tom. Hop in," he said.

"No, thank you. I'd rather walk back. I'll catch up with you all in a bit." I threw my jacket over my shoulder and began to walk down the dirt road. I had a lot to ponder in my heart.

*Whoever else was flying in their cockpit, whatever parted the fog over the runway, it was the crew that used all the resources at hand—and not at hand—to be in a position to capitalize on the breaks they got.*

*The pilot of the Boeing 727 in the next account consulted his crew and an expert on the ground who had worked on the air-*

*plane's certification. It was a case study in cockpit resource management; he considered the inputs and then made his decisions. Nevertheless, in the end he had only his own simulator-sharpened skills to rely on.*

# Gear Door Jam
## *by S. Thomas Peace III*

As we neared Mount Hood, Flight 314 from Salt Lake City to Portland began its descent. We were asked to accelerate to our maximum speed to lead a line of merging aircraft into the arrival pattern for Portland. With the increase in airspeed, the noise in the Boeing 727 cockpit became intense. Suddenly, as we passed through 27,000 feet, there was a loud snapping sound that seemed to come from the very frame of the aircraft itself. It was a sound that I had never heard in all my 28 years of flying. Accompanying the snap was a deep shudder and continuing airframe vibration.

"What—?" The rest of the sentence stuck in my throat. The first officer pointed to the "Doors" light on the landing-gear annunciator panel, which had illuminated. Almost simultaneously, the flight engineer called my attention to his more complete annunciator, which showed that the left main landing-gear door was not in its up and locked position. Indeed, my posterior told me that the door was at least partially open and hanging down into the slipstream, causing the shuddering vibration of the aircraft.

I immediately closed the throttles, extended the speedbrakes and raised the nose of the aircraft to slow the rate of descent and thus decrease the airspeed as rapidly as possible. I wanted to reduce the air pressure against the door to prevent possible damage to it and to prevent, perhaps, even blowing it off the aircraft.

I told the first officer to raise the landing-gear handle from the "off" position, which had removed all hydraulic pressure from the landing-gear system, to the "up" position, which would direct some 3,000 psi to the cylinders that raise the landing gear and its doors. Normally, if the mechanical locks fail, this forces the gear and doors against the up-locks, closing them and extinguishing the warning lights.

This time it did not work. The lights remained illuminated and the shudder continued. The flight engineer had immediately gotten his pilot's operating manual out and turned to the pages concerning landing-gear problems. We could find no procedure in the book to exactly fit this problem.

I asked the flight engineer to check the hydraulic fluid quantity, fearing that we had a leak. It was holding steady. Normally, when a hydraulic line or component fails, fluid loss is almost instantaneous. The fact that we had lost none was good, but also puzzling. If the hydraulics could not raise the door, what had caused the problem?

Just outside the final approach fix, approximately seven miles from the runway, we extended the landing gear. Sure enough, the left main landing gear failed to extend properly. The right main gear and the nose gear extended and locked into place correctly, but the left remained in-transit, neither down nor up.

I asked my first officer to raise the gear. Now the right main and nose gear retracted properly, but the left main gear failed to retract. The flight engineer, his face going pale, called out that the hydraulics were slowly leaking down. My copilot and I grimaced at each other. Quickly, I told him to lower the gear again. As soon as the "good" landing gear was down and locked, the flight engineer turned off the hydraulic system pumps on that system to preserve the remaining fluid.

We made a low pass over the Portland field so that the tower could use binoculars to observe the landing gear and tell us what it saw. The tower reported that the left main landing gear appeared to be only partially extended. A quick mental calculation told me that I only had about 35 minutes of fuel to burn while resolving the problem and then returning to the airport. One plus was that the weather in Portland was good; but this meant that we did not have the fuel reserves that we would have had if the weather had been IFR.

The emergency landing-gear crank was unlikely to free the gear; it only unlocks the gear doors and releases the main gear up-lock so that the gear can fall free under its own weight. We already knew that the gear was partially down. However, we attempted to use the crank anyway; it would not even turn.

The maintenance coordinator found one of his staff who had been working with Boeing during certification of the 727, and put him on the radio. This old hand was a real expert on all of its systems. In a reassuring voice, he said that something similar had happened twice during the test-flights for the 727 and his guess was that the same situation had occurred here. (Later, he turned out to have been exactly right.) He surmised that an actuator involved in the sequencing of the various parts of the landing-gear door during extension and retraction had broken. This caused a portion of the door that should have folded up on gear extension to remain protruding; the wheels had become entangled in that part of the door and could not go up or down.

Our expert suggested that we could speed up and try to blow the door off the aircraft, or dive the plane and pull up sharply, hoping the G load would pull the gear down with enough force to rip the door off, or at least get free of the door.

Could we clean up the aircraft, meaning retract the other landing gear and retract the flaps, and speed up to try to blow the door off? Certainly in such an attempt we would lose the remaining hydraulic fluid in that system. This would mean reduced controllability during the approach and landing, as some flight controls would be adversely affected. I quickly scrapped that plan. The only option left was the dive maneuver.

I told the passengers about the problem and my plan to dive the aircraft. I told them to fasten their seat belts, make sure that there was no loose carry-on baggage that might strike them and check that their seats were in the most upright position and locked. I stressed that they should pay close attention to the flight attendants. Next, to break the tension, I said that they should consider this to be a fun ride on the roller coaster at the county fair. I gave a five-count countdown, then dove nearly 1,000 feet and began an abrupt pull-up. Our cheeks sagged with the G load, but the landing gear did not move. During recovery I pushed the nose over into a mild negative-G maneuver to see if I could float the gear off the door, to no avail. I tried the dive one more time, steeper; no luck. By now, we were down to the fuel situation that required we leave the holding pattern and proceed to the airport.

We had one last aircraft decision to make. Did we want to land with the "good" landing gear extended or try for a belly landing with the "good" gear retracted? If we landed with the "good" gear retracted, the partially extended left gear would create an uneven drag on the runway, possibly causing us to cartwheel out of control. I had, on at least two occasions, practiced landing on just two good landing gear in the simulator after a landing-gear brake fire. Gear-down became the choice. There was no procedure in the book and as far as I knew, no one had ever actually attempted such a landing.

We headed for the shorter runway. It had the desired right crosswind, no nearby obstacles and was wider than standard.

On the downside, it had no glideslope assistance, neither in the form of an ILS nor a VASI. One last runway decision: not to have it foamed by the fire department before we landed. Since we were going to try to use our landing gear, I felt that foam would destroy our braking ability and impair our steering. I wanted a quick stop and foam would cause extended sliding, prolonging our coming to a halt.

Due to a design quirk on the 727, with 20 degrees of spoilers extended any movement of the control wheel would cause a full-up deflection of the "down" wing spoilers and a complete retraction of the "up" wing spoilers. With this technique, we could get the majority of the weight of the aircraft on the good right landing gear and the nosewheel. At our weight, we would be touching down at about 125 knots. I hoped that I could keep the left wing, which had no landing gear, aloft until we had at least slowed to about 90.

We were crossing the Columbia River, and I gave the one-minute warning. At 200 feet on the radio altimeter the flight engineer turned on the hydraulic system. I felt the aircraft lurch as the remaining controls came alive. Fifty feet, right crosswind, right wing down and left rudder to keep the nose straight. Ten feet. Flare. Raise the nose, throttles to idle, don't float! Easy now, get that right main down! Contact! Spoilers to 20! Now lower the nose. Don't let it bang! Hold that left wing up! More right aileron! More aileron! Reverse! Center engine to the stops!

I realized that I was literally standing on the brakes. We were slowing rapidly. A quick glance at the airspeed indicator showed less than 100 knots. I was running out of aileron authority and the left wing settled to the runway. I pulled all of the engines into full reverse and extended all of the spoilers. I removed my left hand from the control column and reached for the nose-gear steering wheel.

I was still standing on the brakes and steering hard right against the wing trying to pull us to the left as it settled on the outboard flaps. I could hear the grinding of metal on concrete as the aluminum of the flaps ground away, leaving the steel flap actuating screw jacks to scrape along the runway giving off a shower of sparks. As the flaps ground away, the nose of the aircraft rose and the nosewheel steering became largely ineffective. We came to a slow stop and as we did, the aircraft rotated slightly left. I could see a fire truck out of my left window and it was spraying foam on the left wing. I later learned that the firemen had seen the shower of sparks from the steel screw jacks and had sprayed foam on the wing as a safety precaution. There was no fire.

We had stopped in less than half the length of the runway, and the nosewheel was less than 12 feet from the centerline. The flaps and leading-edge devices had supported the wing to the extent that the wingtip never touched the ground. All the passengers were accounted for and had suffered not so much as a scratch.

*While the crew of the 727 worked together as they were trained, the pilot of the Cessna 206 in the next story relied on the instructor in the right seat for input, but never thought through what he was doing. Instead of having the advantage of their combined knowledge, the two pilots never quite decided who was in charge.*

# Short Stop
## *by Kent Flowers*

The wind was calm and not a cloud marked the sky as I finished preflighting our Cessna 206. It was a good airplane, but had accumulated the usual number of squawks and glitches that come with a high-time, hard-use machine. I looked forward to the day's flying. My instructor had wangled a charter that involved dropping cargo at several small airports, with a return that evening. He called and asked if I'd like to go along to help build some hours toward my commercial ticket, and I jumped at the chance. I had less than two hours of flying time in the 206 and knew I could use the time to build my proficiency.

After the preflight he offered me the left seat, and I climbed aboard. The first two legs were uneventful. We landed at airports I knew well with wide, long runways. One was an international jetport where standard procedure was to land long, aiming at the middle of a 12,500-foot runway, to avoid a 10-minute taxi among the jetliners to the terminal. The only problem encountered was a 20-knot headwind, which, coupled with a slight fueling delay, had us running about 30 minutes behind schedule. As we departed on the longest leg of the trip, my instructor, a 2,000-hour ATP, told me to "keep it hot and we'll make up some time." He dialed some music on the ADF and relaxed while I played autopilot, dutifully holding altitude, heading and airspeed.

As we approached the small mountain airstrip, he told me that I would not see the runway until he vectored me onto final approach. He pointed to a clump of trees and said, "Keep your speed up and descend to 6,500 (500 agl) over those trees." I pushed the nose down, headed for the trees and started looking for pavement. As we crossed over the trees he told me to start a left turn to final, and I picked up a runway 45 degrees ahead and well in front of me. I lined up and was just reaching to add power when my instructor yelled to chop power. I realized my mistake and saw the real runway much closer and farther to the left. I pulled the nose around to line up on the new heading, pulled the throttle out and hit the flap switch. When the flap indicator reached 20 degrees I shoved in the prop and mixture controls and turned my attention back to lining up on the runway. My instructor yelled to watch the airspeed. Meanwhile, I began to worry about the touchdown.

I pushed the nose down to pick up airspeed, added power to stop the sink, pulled the wheel back to start the flare and heard the stall horn and the wheels bang at the same time; the airplane began to porpoise down the very narrow runway in a crosswind. My instructor said, "Let go and I'll take it around." He added power and took the wheel but the airplane would not fly. He cut the power and yelled, "Hold on," and we kept bouncing, finally stopping near the end of the 4,260-foot strip. And just to make the ride more exciting, we had intermittent left brake effectiveness (we both knew about this weakness before takeoff), which made the airplane swerve from side to side during the roll-out. It was as close as I would ever want to come to an out-of-control landing.

As we congratulated ourselves on living through another landing, we noticed one other detail—the flaps were extended to a full 40 degrees, hence our unusually steep final approach and our inability to go around after the first bounce.

I spent the rest of the trip thinking about that landing. Was it poor crew communication? Was I really flying the approach, or was I just acting as the voice-actuated autopilot for the man in the right seat? Was it my low time in type and unfamiliarity with the landing characteristics of this airplane? Was it our rushed, hot approach, lining up on the wrong "runway"? Or was it the broken spring in the flap switch that I forgot in the flurry of activity, allowing full flap deployment?

Whatever the reason (or combination of reasons), I made some solid resolutions at the next stop, as I pulled alfalfa branches from the landing gear and wiped the telltale grass stains off the wheel pants. First, that when flying with another pilot I will have a clear understanding of who is flying the airplane and who is doing what during the flight. I will fly full patterns to allow time for checklists and a good look at the correct runway, just to be sure. I will insist on a complete checkout in any airplane I fly. I will abandon any approach that I am not totally satisfied will result in a safe landing, and I will start my go-around before the first bounce. And I will never tell a single soul about that botched landing.

# 9

# Flying on the Edge, Pushing the Limits

Flying often pushes pilots into situations where they are forced to perform beyond their past experience or to make demands on their aircraft that push the performance envelope further than the designers had intended.

Smart pilots understand that they've exceeded their limits and been allowed a second chance. Not-so-smart pilots use the experience to raise their outside limits. Eventually they get into more serious trouble. The hard lessons should never have to be taught more than once. Reducing the margins also reduces the chances of a safe outcome. In analyzing an accident after it happens, there is often a chain of circumstances that led to the final, and at that point, almost inevitable outcome.

In the confession that follows, a pilot who has not made a night landing in more than ten years finds himself taking off for a destination that isn't particularly hospitable for night operations and counting on tailwinds to get him there before dark. Even if nothing else had gone wrong he would have been cutting it close. But other things did go wrong. . . .

## The Sun Also Sets
### by Dan Froseth

It was not the proverbial dark and stormy night, but judging from my sense of foreboding, it might as well have been. Instead, it was a clear and crisp early evening in January when we departed Sacramento Municipal Airport aboard our faithful

Cessna 172. We anticipated tailwinds, which would get us home just as the sun set.

The takeoff and climb were normal, and my wife and I admired the enormous view of the Sacramento Valley. We relaxed and looked forward to what should have been a short and uneventful flight to our home base in Redding, a little more than an hour to the north. Full fuel, clear skies, familiar terrain, experienced pilot—what on earth could go wrong?

Fourteen-thousand-foot Mount Shasta dominated the distant horizon, while familiar, less imposing landmarks slid by in the flatlands below. But as I navigated I began to notice that the landmarks weren't sliding by as fast as they should have been. Our tailwind had turned into a blustery headwind.

Off to the west, the sun was determined to set in its usual location at its predicted time for this January day. But in the eyes of this pilot, the sun soon appeared to be getting ahead of schedule—my schedule, that is—not cooperating at all with the slow progress of our Skyhawk against increasing headwinds. A night landing was definitely not part of our agenda, since I had not performed one in more than 10 years. If the winds and the cosmos would just settle down and cooperate, I had every intention of happily planting the airplane on a runway before it got dark.

We didn't need any more complications, so, naturally, we got one. A little red light on the panel began to seek our attention by flickering "high voltage." I reset the master switches and checked all of the circuit breakers and the pesky little light went off. For the time being.

We were approaching Red Bluff Airport, within 15 minutes of our destination, when the light came back on and stayed on. The mental struggle between "Press on" and "Play it safe and land" began:

"Land."

"But we're almost home."

"Land."

"But we won't have any ground transportation."

"Land."

"But look, the warning light is fading out."

"Maybe."

We pressed on. There was still adequate daylight and we were determined to spend the night at our home base. Off to the east, along the base of the foothills, a vehicle with emergency lights flashing was racing toward Red Bluff. It paralleled our course and arrived there before we did, confirming my diagnosis of deteriorating groundspeed.

Our repeated calls to Red Bluff were met with silence—the radio was dead. Fifteen minutes to home had stretched into 25.

The clear, friendly skies of northern California were rapidly fading to darkness, and our Cessna had experienced a complete electrical system failure. We had no radio, no instrument lights, no flashlight, no landing lights, no flaps and no recent night-landing experience.

"Relax," I said to myself.

"Easy for you to say," self said back.

What are the options, I wondered, aside from waiting for the sun to return? Not many. Benton Field was slightly more than 2,000 feet long, in daylight. But with the sun making a thoughtless and seemingly premature departure, Runway 33 appeared to have shrunk to the length of a football field.

Redding Municipal had longer runways and would be a more hospitable night-landing site for a 172 and pilot fully equipped for a daylight flight. Unfortunately, we had no radio capable of contacting the nice people in the control tower. We felt somehow that we couldn't add our unlighted, uncommunicative airplane to their busy air traffic situation.

We went back to plan A. Relax, fly the airplane. *Feel* the airplane would be more accurate, because the instruments had gradually disappeared in the darkness of the cockpit on this moonless night. I looked out over the lights of the city and tried to remember what the runway lights should look like on final approach.

With no electric flaps available, it would be a flat final. I listened very carefully to the sounds of the engine and the air rushing by the airplane and tried to use all my senses (something I had obviously failed to do earlier) to search for the elusive airport beacon.

With the beacon in sight we started down, feeling our way as if blindfolded. I set up a long final, carefully watching for traffic, since it would have trouble seeing me. I eased the airplane into the landing slot, doing my best to estimate my altitude, airspeed and rate of descent in reference to the double row of white lights outlining the runway ahead.

Then the green runway-threshold lights slipped by underneath us. Bang! We were on the runway lucky and grateful. We slowed to a midfield exit and endured the curious stares of other pilots operating airplanes with fancy options like taxi and landing lights. We taxied very slowly to our tie-down.

The next day we learned the cause of the electrical system failure: the alternator had partially disintegrated, shorted out and quickly drained the battery of all its power.

A few questions emerged from this flying fiasco. Is it possible that the pace of the setting sun is accelerated by headwinds? Not likely. Is it a good idea to plan one's arrival near dusk, when not fully prepared for a night landing? Definitely not. And what about not taking seriously the faint flicker of a warning light when there is adequate time and daylight to deal with it? Don't ask.

✈ ✈ ✈

*Pressing on in the face of increasing arguments against it is one of the most common scenarios in aviation accidents. A red warning light flashed on his panel as he overflew an airport while it was still daylight. But it wasn't home and that's where he wanted to be. First we deny there's anything wrong, and then we believe in miracles. "Maybe it'll fix itself," we argue as we press on and almost inevitably exacerbate the level of the emergency.*

*The problems aren't always with the airplane or equipment. Pilots, too, have been known to malfunction. You wouldn't take off with a critical instrument acting up, but in the following case, the pilot first denies there's anything wrong and then convinces himself that coffee will fix him right up. Wrong.*

# Party Animal
### by Al Starner

It's a party and I love parties. The drinks are on the house and we're having a ball. We have just completed the Army primary fixed-wing aviator course at Fort Rucker, Alabama, and tonight we're celebrating graduation with our flight instructors. It's fun to rehash stories of groundloops and student pilots who got lost on the first night cross-country flights.

Tomorrow is our first day of tactical instruction, and as the party wears on quite a few of the students and instructors leave early. What a bunch of party poopers. I order another drink. My instructor has switched to a soft drink, and I wonder if he has a bad stomach. Last call is announced at midnight. I finish my drink and go home.

The alarm clock rings 30 seconds later, or so it seems; it is actually five a.m. As I shower and climb into my flight suit I notice that things are still blurry and my head throbs. A few cups of coffee should straighten me out, I think. I catch a few winks on the bus to Lowe Army Airfield, but I still feel woozy. Oh well, the ceiling looks pretty low and we will probably be grounded. But we aren't.

After preflight and run-up of our Cessna 0-1 Bird Dogs, we take off and head for the tactical training area. I'm Gold 33 and flying solo. The aircraft takes forever to leave the runway; then I see why—I've taken off with the carb heat on. My instructor would have chewed me out. I shrug it off and concentrate on navigating to our rendezvous area.

The usually familiar terrain is confusing and I have trouble orienting myself. After flying in circles for a while, I recognize a drive-in movie screen in Enterprise, Alabama, and use it as a reference point. While scanning the instruments, I notice that the auxiliary fuel pump is on. Another screw-up. I must have left it on after takeoff.

Suddenly my radio comes alive: "Gold 33, proceed to strip 101." During primary training we reviewed all the tactical strips, and I find 101 with little trouble. I see the two other airplanes orbiting the strip for the prelanding recon. As I work my airplane into its recon pattern, my instructor says sarcastically over the radio, "Glad you can join us. Gold 33, you will be third to land."

The strip isn't difficult, but a large oak stands at one end of the runway and the landing approach has to be steep to clear the obstacle.

As I start my prelanding checklist, I'm horrified to see that I haven't switched fuel tanks. I've nearly emptied one side. I switch tanks while the second student makes his approach.

While descending to enter the downwind, I see the second airplane touch down. The student pilot who landed has real problems with short fields, so the strip can't be tough if he got in. As I close the throttle, I see that I've lost altitude on the downwind; I'll have to turn base early so I can keep enough altitude for a steep approach. I cheat a little and drop flaps early to slow the airplane. Turning final I still have too much airspeed, so I drop full flaps to slow down. Suddenly it's clear that I'm going to touch down halfway down the runway, with insufficient room left to stop the airplane. I break off the landing and go around. The engine seems sluggish when I add power. Carb heat. Again, I forgot to move the knob to its proper setting. The airplane staggers along at a high angle of attack, with full flaps and reduced power. It nearly stalls. During my recovery I brush the treetops at the end of the runway, which really shakes me up. I regain altitude and turn downwind again. I try to concentrate. But the second landing attempt is as bad as the first. I radio my instructor, who's been watching from the dual airplane. "My fuel is getting low," I say.

"Return to Lowe," he answers.

Back in the classroom, my instructor arrives. He quickly critiques the other students and dismisses them. He then turns to me.

"Low on fuel, huh?" he asks. "We both know what your problem is today." I nod and start to speak. "Shut your mouth," he says. "You shouldn't have been flying. We both know I can wash you out of flight school right now if I choose to."

I nod.

"I'm going to ask you one question. What did you learn today?"

My answer is simple: "Never mix booze and airplanes."

✈ ✈ ✈

*The "Party Animal" chose to fly although he'd had too much to drink the night before. In the next confession, the pilot again should have known he was pushing himself—perhaps beyond his personal limits—when he elected to fly with a head cold. He kept telling himself it wasn't as bad as he knew it was. And it nearly got worse. . . .*

# Facing Pressure
## by Lt. William R. Shivell

It happened during a major pre-deployment readiness exercise. The battle group was off the coast of San Diego, making ready for a simulated "war." The exercises, although grueling in their intensity and duration, offer some of the best training Navy pilots ever get, as well as the chance to bag lots of hours and traps [carrier landings]. Naturally, I wanted my share. During the first week we participated in scheduled exercises; during the second week we were slated for "war"—which meant round-the-clock alerts, little sleep and lots of flying.

Days One—Three: Uneventful. Day Four: I awoke with a sore throat and dreaded its implications. A sore throat alone was tolerable, but if a head cold was on the way, in two to three days—just in time for "war week"—I would have a stuffed nose and blocked ears. "Some of the best flying in the cycle, and I'm going to be MED down," I thought.

Day Five: The sore throat persisted, but the cold had not progressed, so I flew the day's mission. Day Six: I awoke with some congestion, but it seemed to clear while I took my morning shower. Again, I flew. During that flight I began to feel some discomfort in the bridge of my nose. It felt as if my oxygen mask was on too tight. The discomfort continued throughout the flight, but when the pain disappeared after getting back on deck, I quickly forgot about it.

Day Seven: I was scheduled for the first flight, a surface surveillance mission requiring several climbs and descents to simulate the sighting of surface contacts from altitude followed by descents to investigate those contacts. I awoke at 0515 with a full-on, stuffed-up head but took a shower and went to the brief. My RIO (Radar Intercept Officer) asked whether I was up to the flight. I said I could hack it. The congestion had cleared somewhat, although I was blowing my nose constantly. Even as we manned the jet, I had to remove my oxygen mask several times to blow my nose. My RIO warned me that it wasn't worth flying with a cold. I gave his warning some thought as the "yellow shirts" (flight-deck taxi directors who wear yellow gear for identification) taxied us to the catapult, but it wasn't enough to stop me. Like many pilots, military and civilian, I was feeling quite invincible.

We launched, climbed to 20,000 feet and began a search of our sector. About 20 minutes into the flight, the pain in the bridge of my nose returned. "Guess the PRs (Parachute Riggers, the guys who take care of all the flight gear) were working on my mask," I rationalized. We found some contacts, and once we overflew them at altitude, started an aggressive descent. Passing through 8,000 feet, we were in a 30-degree dive. I concentrated on clearing my ears and setting up for photo runs on the contacts.

Passing through 5,000 feet, the pain in my nose had spread to my eyebrows. Concerned, I eased our rate of descent. At 3,000 feet, the pain was no longer an afterthought. I leveled off. My RIO said that we had to keep descending if we were going to make the first pass. I told him that something was wrong, yet I couldn't define exactly what that something was. He suggested that we make a shallow descent and get the contacts on the next pass.

The pain worsened during the shallow descent. At 1,500 feet, it felt as though ice picks were being driven into my eyebrows. Digging my heels into the cockpit floor, I tried, near tears, to tough it out. Finally, at 1,000 feet, I began a zoom climb-out to 5,000 feet. I told my RIO I had a major problem.

Now I knew what was happening to my sinuses, and why. I had a blockage, which prevented pressure from being equalized. That I should have known better than to fly with a cold did not cross my mind. To land, we would have to descend to 60 feet, the elevation of the flight deck; but I also knew I couldn't physiologically withstand going below 1,000 feet.

We decided to make shallow descents in 1,000-foot increments to 3,000 feet (VFR holding altitude at the ship) and remain at that altitude until we had to land—hoping the pressure in my sinuses would by then equalize enough so that I could withstand the final descent. We had more than an hour to wait but were looking good on fuel. Working on down to 3,000 feet, the sharp pain subsided into a dull ache. We continued the mission at 3,000 feet until it was time to report overhead.

Then nature turned against us. The weather went bad enough to require a "Case II" approach—the Navy version of an IFR approach to circling minimums. This required me to initiate a climb-out from the more physically comfortable altitude of 3,000 feet to 8,000 feet, and hold there. Foolishly, I didn't want to attract attention to myself, so I did not request special handling at 3,000 feet. I was resolved to concentrate on flying the airplane and prepare myself for what was coming—pain, which only intensified as the approach progressed. The more severe the pain, the more I concentrated on flying. When the jet shuddered to a merciful halt on the flight deck, I uttered, "Thank God!"

As we taxied on deck, my only thought was to get to medical, and quickly. Upon seeing the flight surgeon, I said, "Doc, I've really —— up!"

According to the doctor, that was an understatement. The X-rays showed it all: I'd suffered a hemorrhagic barotrauma. He explained that a sinus was just an air cavity that needs to equalize pressure through small passages. When someone has a cold, these passages can be blocked by swelling or mucus. In a climb, air needs to escape to equalize pressure. In a descent, air needs to reenter the sinus. If the passages are blocked, a vacuum will form; this is what is known as a "sinus squeeze." The greater the ambient air pressure, the worse the pain. This will usually result in a hemorrhage of the sinus. That's what happened to me.

The flight surgeon told me I was grounded for a minimum of 30 days—if there were no complications. Another doctor on board said he thought I would never fly again. Fortunately, my sinuses didn't actually collapse, or else surgery would have been required. I was immediately given antibiotics to prevent infection, decongestants and painkillers (the "ice picks" now felt as if I'd received a general kick in the face). I sat out the rest of the "war" either in my stateroom or at the duty desk, wondering if I had just had my last flight as a naval aviator.

To most people, a cold is just an inconvenience; but for someone flying or riding in an airplane, it's something that can destroy you. The cabin pressure of an F-14 is maintained at 8,000 feet, the same as any airliner or light airplane flying at typical cruising altitudes. What happened to me could have happened anytime, in any airplane.

Your level of maturity can determine whether a cold will ground you for a few days or permanently. I was lucky. Instead of being grounded permanently, I only missed out on about seven traps, my first Sidewinder missile shoot and a month's worth of flying. Both the pain and the worry about my future in naval aviation lasted only for a few weeks. It could have been much worse: surgery, convalescence, reassignment to another squadron, killing my RIO and myself. In the end I realized I had paid a small price for learning an important lesson: No one—not even a pilot—is invincible.

*"Like many pilots, military and civilian, I was feeling quite invincible," the military fighter jock confessed. His feelings almost grounded him—permanently. But the feeling of invincibility—being able to handle any situation—can be insidious. In the following incident, when a layer of fog and cloud over a river separates the pilot of an Aeronca Champ (with minimal instruments) from his home base ten miles away, he decides to hop over the fog. No big deal. He can handle it. But can his Champ?*

# Fog Bound
### by Richard Mazziotti

The return trip to New Jersey from a weekend in the Chesapeake Bay area was bumpy, hot and humid. But as a new tail-dragger pilot, I was enjoying the trip in my recently acquired

Aeronca Champ. Considering that it was almost as old as I, the airplane was in remarkably good shape. Everywhere I went some grizzled hanger-on would come up to me at the gas pumps or tie-down area to say, "Oh yeah, I've got 200 hours in one of these," indicating that (one) they were taildragger pilots and proud of it, and (two) they all had more time than I did. The airplane was equipped with a compass, airspeed and oil-pressure gauges, and a great bloody Spitfire-type turn-and-bank indicator whose action lagged so greatly that it was virtually useless.

Over eastern Pennsylvania a gradual purpling of the sky in front of me showed that the forecast of possible thundershowers was accurate. I'm a devout coward by nature, so my plan was to continue to edge around the growing thunderstorm to see if I could outrun it to my home airport, just a few miles away, unless I felt that first big bump of turbulence.

Duly came the bump, a smell of wet air, a few big drops on the windscreen, and I turned right toward nearby Blairstown Airport. The landing at Blairstown was as uneventful as any landing by a novice in a taildragger in a gusty crosswind ever is. Suffice it to say the airport hangers-on were hangaring on with more than usual enthusiasm as I skittered and hopped about in the general vicinity of the runway. I taxied into the high grass and tied down just in time to get drenched running for the office. Like most late-afternoon summer thunderstorms in New Jersey, this one cleared the area about seven p.m., leaving a few low clouds. Since home base, Sky Manor, at Pittstown, was just across the Delaware River, about 10 miles away, I did a fast preflight, propped the Champ in the now-quiet air, snuggled into the front seat and took off, heading for home.

The leftover scud was lower than it looked. I leveled at about 1,000 feet agl. Ahead was a solid wall of low mist hanging in the river valley. But the pillows of fog and cloud were staying over the river, and my home airport was just beyond the other side. I decided to hop over the fog, quite sure that I would make ground contact immediately on the other side.

Suddenly my world went gray. The Champ's nose and wingtips disappeared as I punched into the now-solid wall of clouds. What was this? I had intended to just slide through that little bit of vapor curling above the fog bank. Why wasn't I coming right out the other side? What was I going to do now? I began to realize that I was in deep trouble and getting deeper.

With no effective turn-and-bank, no navigation instruments and a firm belief that I could not remain head-up and wheels-down for long, I began what I hoped was a shallow 180-degree turn to my left. An occasional glimpse of the ground was enough to keep vertigo at bay, and after what seemed to be a long, long time, I broke out of the cloud. In that short period the clouds and fog had started to coalesce on the back of the mini cold front. While for the moment I was flying upright and keeping clear of the dirt, I was now in danger of getting lost. Successive glimpses of a road, powerline, lake and my lap-clutched sectional finally produced, under my left wing, the airport I had just left. An uneventful landing ended the evening's adventures.

Looking back, I can see that what might have been the real problem was one that would never be listed in an accident report. You see, I was newly instrument-rated. With a reasonable amount of naivete, I knew I could face the occasional cloud without difficulty. What I had neglected to take into account was that even though I was instrument-rated, the airplane I was in was not. The accident report would have read, "Pilot error . . . continued VFR flight into IMC." What a way for an instrument-rated pilot to go.

✈ ✈ ✈

*Doing a 180 finally got him out of the fog, but then he nearly found himself lost above a solid layer with night approaching and no navigation and gyro instruments. Having exceeded the airplane's limits, he was fast approaching his own.*

*Knowing your limitations is important. Not exceeding them is even more critical. As we develop and experience different situations, our skills increase and our limits are stretched further and further. Many times it's while we're still learning that we slip outside the boundaries. A helicopter pilot out on a first unsupervised solo flight learns—the hard way—that there are still limits to what he can do.*

# Skid Marks
## *by Robert Jackson*

My first "demo" flight in a Robinson R22 convinced me that I had to add a helicopter rating to my private certificate. Now, several months later, I was on my first unsupervised solo flight.

The sky was blue, the wind calm. Having previously been rated in single- and multiengine airplanes as well as gliders, I was familiar with that exhilarating mix of fear and excitement that students feel when first leaving the airport environs on their own. But this flight was different. I had done well in my training and felt confident of my abilities. The excitement I felt came from the unaccustomed nearness of the ground. Never before had I the option of landing in one of the fields below me, without the landing being unplanned.

A glance at my watch confirmed that it was about time to return to the airport, but I was having so much fun that I decided to make a practice approach to a field in front of me before calling it a day. By the time I had turned final, I had decided to land. That was my first mistake.

Although I had made off-airport landings with my instructor before, I had no authorization to land in the field I had chosen. I did, at least, have the presence of mind to pick a spot that was free of crops. It was a patch of black earth in a sea of green and easy to concentrate on as I lowered the collective, eased the cyclic back and set the helicopter down.

Impressed with my smooth landing, I smiled and opened the door. My skids were only lightly pressed into the dirt. Then I was seized by my second stupid idea of the day. Why not come around for another landing and see if I could put the skids in the exact same place?

I closed the door, put my hands and feet on the controls, looked out in front of me, and lifted the aircraft off the ground. I was two feet in the air when the door popped open. I had not closed it properly before lift-off. An experienced pilot would have calmly placed the craft back on the ground, lowered the collective fully, put the collective friction control on, and then closed the door. An inexperienced but smart pilot would have done the same. I was neither.

Taking my left hand off the collective, I reached across my body to grab at the door latch. As I did, I took my eyes off the field in front of me and glanced down through the door at the field below. It was moving forward and to the left, which could only mean that I was drifting backward and to the right. My grab for the door had caused me to put a slight side pressure on the cyclic, which upset the delicate balance of the helicopter. The result was uncontrolled contact of my right skid with the furrow behind me.

I raced to get my left hand back on the collective. Having read the safety notices put out by the aircraft's manufacturer, I knew there was no way to stop the imminent dynamic roll-over with cyclic movement alone. Even with full left cyclic applied, the aircraft would continue to roll to the right, unless I could free the skid or lower the collective. I had time to do neither.

Looking straight ahead, I saw the horizon tilt and I felt the helicopter roll. I had enough time for one thought: this shouldn't be happening. The violence that resulted from the meeting of the main rotor with the farmer's field blocked out the next few seconds from my memory. I recall lying on my side listening to the shriek of the low-rpm warning buzzer for a few seconds, before reaching over to turn off the master switch. I managed to release my seatbelt and exit the remains of the aircraft. I climbed out the left door, but stepping out the front would have been just as easy. The Plexiglas windshield was scattered about the field in tiny pieces, along with several other parts of the helicopter. A brief check confirmed that I was not seriously hurt, aside from my pride. For a moment, I could only stand and look at the result of my stupidity.

Finally, I began walking through the field to a farmer's house a few hundred yards away. As I walked, I had plenty of time to reflect on my accident. Did my flight instructors adequately prepare me for solo flight? Was the helicopter just too squirrelly for low-time pilots? The conclusions I came to have not changed.

Like all helicopters, the R22 that I flew that day was perfectly safe when flown within its limits, as well as the pilot's limits. But helicopters are less forgiving than airplanes of a pilot's overconfidence.

✈ ✈ ✈

*Part of being a student involves pushing out the envelope. Gently testing the limits. The student knew he wasn't supposed to land in the field, but his confidence got the better of him. When he did set down, he put into motion the scenario that played out with the Robinson R22 beating itself to death on the ground.*

*Overconfidence was riding along with a Cessna 150 pilot when he learned about high-density altitude, mixture control, the airplane's performance limits and his own capabilities.*

# You Can Be Too Rich
### *by Dean Thomas*

Beginning pilots often hear lore from their more experienced peers that they hope are of dubious veracity. One thing I heard was, "In your first 100 hours of flight time you will do something that will scare you to death, if it doesn't kill you."

This little proverb came true for me one hot August day near the valley of San Bernardino, California. The day was partially smog-free and presented an uncommonly glorious opportunity to fly. I decided to take a friend on a short one-hour sight-seeing trip over Big Bear Lake, elevation 6,750 feet.

The airplane was a Cessna 150, equipped with the standard Continental O-200 engine, onto which was bolted a cruise propeller. (A cruise prop, I later learned, can reduce the engine rpm by as much as nine percent at full throttle, as opposed to a climb prop, which allows the engine to develop its full-rated rpm.)

Most of my 85 hours had been spent in this little airplane, and I was confident of its abilities and my own as we took off from the valley. While we climbed through Cajon Pass (which means "coffin" in Spanish), I gradually began leaning the mixture.

Just before reaching Big Bear Lake, I had a sudden attack of bravado and decided to land the 150 at Big Bear Airport. My instructor had told me that it was not a good idea to take the 150 to Big Bear, but I thought I could do it.

"I've had this thing above Big Bear several times before," I reasoned. "How hard can it be to land it?" The fact that those other times I had not had a passenger on board, nor had the days been as hot, seemed to slip my mind.

Unable to raise anyone on Big Bear Unicom and failing to see the wind sock, I decided to land on Runway 25, which points toward the lake. During my approach, I did all the things I had been taught to do down in the valley. I applied carb heat, lowered flaps, and made sure my mixture was full rich as I pulled back on the throttle.

On final approach I noticed that the ground seemed to be slipping by a little faster than normal, and I searched for the sock once more. I spotted it at midfield, fluttering toward the lake and indicating about a five-knot tailwind. A couple of hundred yards before the threshold I decided I would abort the landing and come back in on Runway 7.

I announced my intentions on the unicom but didn't apply full throttle right away. In another fit of stupidity, I decided I wanted to impress my passenger with my flying skills, so I made a low, slow pass over the runway before heading out over the lake. Most of the runway had disappeared behind us before I finally turned off carb heat, applied full throttle and raised the flaps.

The engine surged briefly but then suddenly died into a mushy idle as the end of the runway slipped away behind our tail.

After my initial disbelief, I realized we were too low to attempt a 180-degree turn and frantically searched for a nonexistent landing site along the fast-approaching tree-lined shore of the lake.

I jammed the throttle to the wall, but the engine responded gut-wrenchingly slowly, and we continued to sink.

I quickly checked the flaps, carb heat and mixture. It suddenly dawned on me that I had put the mixture in the full-rich position.

When I yanked the mixture back out to the lean position, where it had been before my approach, the engine sputtered and kicked back to life. With my heart wanting to leap out of my chest and the cold-looking water now close ahead, I fought the urge to pull back on the yoke as the airspeed hovered just above stall.

With white knuckles, I nursed the airplane low and straight over the lake and listened for the dreaded sound of the stall warning horn. When I finally was able to bring the 150 to a safer altitude and felt somewhat able to control my voice, I announced to my petrified passenger that we were going to make it. My newly enlightened, adrenalin-cleared mind enabled me to make my first intelligent decision of the day, and we headed for home.

"The day you stop learning is the day you become a dangerous pilot," is another proverb that I now hold true. That day I learned how the combination of poor judgment, high density altitude and improper mixture control can kill an overconfident pilot.

Fortunately, I'm still here to say that I haven't stopped learning, but these days I try to take my lessons a little more conservatively.

✈ ✈ ✈

*"My instructor had told me that it was not a good idea to take the 150 to Big Bear, but I thought I could do it."* Instructors do

*have experience they draw from, and generally, student pilots should heed what they say.*

*In the following adventure an instructor and his student both learn a valuable—and fortunately this time—inexpensive lesson.*

# Mixture Madness
### by Chuck Diver

Although it has been 15 years since my primary flight training, my lessons remain very clear in my memory. One in particular returns every time I see Spanish moss. My flight instructor was very good at making lasting impressions from training situations that seemed routine. On this particular flight, I think we both learned a thing or two.

We had flown the Cessna 150 from Jacksonville, Florida, to Waycross, Georgia, and were on the return leg when my instructor decided to simulate an engine failure. My navigation was generally good but sometimes I would fail to follow our flight path on the chart as closely as he would prefer, so to make his point, he waited until I was overdue for a navigation fix, and out came the mixture control.

I began reciting the list of immediate steps to attempt to restart the "failed" engine while trading airspeed for altitude, and having found the inflight emergency checklist, verified that I had done all that I could do, without pushing the mixture control back in. We were at 5,500 feet when the engine "failure" occurred, so I had about four and a half minutes to find and select the best available landing spot within a four-mile radius. Having established the best-glide airspeed, and simulating a radio frequency change to 121.5 and 7700 on the transponder, I glanced at my instructor, and the expression on his face told me this was not the end of the drill.

He just sat there, stonefaced, and said wryly, "Where you gonna put it?" As the propeller slowed to a complete stop, I tried to focus on the folded sectional chart, looking for an answer to the question that was reverberating in my brain. When he unplugged my headset from the radio I knew he was up to something, and I could almost hear him talking to someone, but I was too busy to listen. "Flaps 10," I said, not having an answer to his question.

I knew we were near Folkston, Georgia, and I could see several inviting roads and a highway, and there were fields that, from this altitude, looked flat. "The highway has too much traffic on it so I think that field is my best bet," I said, pointing to the one that was long in the upwind direction. "Do a 360 to the left," he said. "Okay" was all I could come up with. We were passing through 4,000 feet and I knew there was time to take a look behind us. Not seeing any better options, I said, "I still think that field is the best choice." We were now passing through 1,500 feet, and I could feel the yoke beginning to sweat. "Do a 180 to the right," he said, "and look below us." Again I could see him talking to someone on the radio but I could neither hear nor had time to try to listen. Halfway through my turn, as I looked past the wingtip, appeared, like a vision, an airport!

In one split second the lesson hit home: If you always know exactly where you are you will know what options are available. We had been above the Davis Airport when the engine "failed" and I had not seen it because I was looking around for a place to land while the airport was hiding behind the trees and the tip of my wing. With 800 feet and about 30 seconds between us and the tops of the pine trees, I set up our final approach as if we were actually going to land on the runway. My mind was racing through what I had to do, what I had done, and whether we were going to clear the tops of the trees between us and the threshold. What had been a leisurely descent had gradually become a rush of Georgia pines. I knew any second I would see him push in the mixture control signaling the end of the drill.

"Watch your airspeed" was not what I had expected to hear him say, as he took off his David Clark headset and turned up the speaker volume for me to hear our clearance to land, "and make sure you don't overshoot." Overshoot? I thought to myself, I don't think we are going to make the runway . . . "And plan on a full stop," he interjected. I knew I would have to thread the needle to pass this test. The windsock indicated almost no crosswind, but it felt as if there was, and I realized that I was out of trim; what else had I overlooked?

As we passed through 600 feet my concern about not making the runway became a realization that we might be too high to land on such a short runway, and once again, my instructor was right. Instead of remaining in the same relative position in the windshield, the runway numbers were beginning to slide below the nose. "Flaps 40," I said, as I held the switch and the nose down, feeling the drag decelerating us like an air brake. As the tops of the trees got close I realized that we were too high to land safely and there was only one remaining option: Start the engine and try it all over again.

"If this had been a real emergency you would be dead meat," he said, looking disgruntled. "Start the engine," he barked, as he pushed in the mixture control. I watched the threshold slipping beneath us as we passed through 400 feet. I was embarrassed, depressed and relieved all at the same time, but only too glad to restart the engine.

The restart procedure was straightforward, following the memorized checklist, but all I got was *ra ra ra ra ra ra ra*. I searched for something I had overlooked but everything was as it should have been. *Ra ra ra ra ra ra.* Time seemed to slow down as I watched our hands working the throttle, mixture, carburetor, heat, magnetos, ignition and primer, like a duet of desperate but measured actions. The altimeter read 150 feet and we had maybe seven seconds before we would be on what was left of the runway, and to the eye, it did not look as though there was enough room to stop before we reached the fence. For some reason, I remember looking up at the trees that lined the runway and seeing large gray clumps of Spanish moss hanging, blowing in the wind. Flying the airplane was a subconscious function during this time, thanks to the training of my instructor who for the moment was totally occupied with starting a balking engine.

At last, the engine came to life with a roar and we began pulling ourselves back up and away from the runway. "Flaps 10," said my instructor, as I reached for the switch. As we raced toward the trees, we just sat there watching and privately calculating by how much we would clear the tops. There was that Spanish moss again. We cleared the trees by only a few feet, but we had made it. "Flaps up," I said, as our airspeed increased. We both sat there, quietly staring through the windshield, listening to the engine, watching the trees passing farther below us, taking in the beautiful blue sky, and collectively breathing a deep sign of relief.

"Do you want to try that again?" asked my instructor. As I turned to look at him, he adjusted his headset over his eyes like a blindfold, and we both laughed out loud. The release of tension said, in effect, we had both learned enough for one day.

*"If this had been a real emergency you would be dead meat," the instructor said, and then it very nearly became one when the engine balked. Instructors aren't invincible, but often it's helpful to have one along. In the following adventure, a pilot decides he can act as the instructor and lets his buddy ride in the left seat while he flies from the right. Instructors do it all the time—how difficult can it really be?*

# When Right Is Wrong
### *by Robert Waldrum*

I had read stories about 100-hour pilots who believed they knew what they were doing and then got themselves into trouble—but I never thought I would be one of them.

Wanting to introduce a friend of mine to flying, I asked him out to the airport one day after work. I had trained at an FBO based at Palomar Airport in Carlsbad, California, and while in training had flown mostly Piper Archers and Warriors, although I had managed to also log a few hours in a Cessna 152. After obtaining my private pilot license, I continued to rent Archers and Warriors from the same FBO.

Although I hadn't flown the Cessna 152 for about six months, I remembered soloing in the little airplane and thought it would be a nice airplane to use for my friend's introductory flight. To reacquaint myself with the 152, I flew it briefly just a few days before my friend's "introduction" to flying.

My friend arrived at the airport just as I was finishing the walk-around. Without a second thought I suggested that he sit in the left seat—where I thought it would be easier for him to try his hand at flying.

Although the motions of starting the engine from the right seat felt strange to me, I continued on, thinking I would eventually shake the uneasy feeling. After all, I was a pilot with more than 100 hours!

From the right seat, maneuvering the 152 on the taxiway proved difficult for me. I seemed to lurch the airplane around the turns as if I had two left feet. Arriving at the run-up area and jerking the little Cessna around into the wind, I started to have serious doubts about flying from the right seat. I glanced at my friend, who was anxiously anticipating getting his hands on the controls once we took off. What a disappointment it would be for him to stop now, I thought. This, plus the fact that we had only about one hour of sunlight left, made me continue.

Completing the runup, I indicated our departure heading to the tower and we were cleared for takeoff. From the right seat I pushed the throttle to the wall and tried to maintain centerline during the takeoff roll. As the 152 built up speed, I became very anxious; sitting on the "wrong" side of the airplane, I felt disoriented at the controls. Familiar positions were now reversed: instead of my right hand on the throttle and left hand on the wheel, the left hand was now on the throttle and right hand on the wheel. Everything felt wrong, but I continued the flight.

I rotated the 152 without incident, and after establishing a positive rate of climb, we headed out to one of my old practice areas. During climb-out I began to feel more relaxed. Flying the Cessna was easier than moving it around on the ground. Or so it seemed.

Leveling off at 3,500 feet, I put my friend through some of the normal maneuvers a student does on the first few flights. After about 45 minutes, we headed back to Palomar. Entering the pattern—right downwind for Runway 24—I again began to feel disoriented. Landing reference points did not appear the same from the right seat as they did from the left. I also had the sensation of drifting over the runway, because I could look down and get an unobstructed view of it—unlike the view of a right-hand pattern from the left seat. Abeam the numbers, I adjusted power, applied flaps and trimmed for approach speed. As I started my turn to base, I felt I didn't know where I was or what I was doing.

The 10-knot wind off the right side of the cowling was pushing us left. I put the Cessna into a slight slip to compensate, which of course increased my rate of descent. On short final, the VASI went from red over white to red over red, bringing to mind the rhyme my instructor once taught me: *White over white, you'll fly all night; red over white, you're all right; red over red, you're dead.* Adrenaline raced through my body.

Seeing red over red, I "added" power, all right with what would have been a normal thrusting motion with my right hand—had my right hand been in its usual position on the throttle. But instead, as I instinctively pushed with my right hand, I was actually pushing the wheel. The Cessna dove for the ground. In a real panic now, and wanting to pull back on the yoke to regain some altitude, my left hand automatically yanked back. But my left hand was not on the wheel—it was on the throttle. As a result, I pulled the throttle back so hard I cut power completely.

Ten feet above the ground and some 50 yards from the threshold with the engine at idle, I had to mentally force myself to push in with my left hand, and pull back with my right hand. I finally began to establish a positive rate of climb, but now I had too much nose-up trim applied. I had to forcibly hold the nose down with my left hand to keep from stalling.

While I was seesawing up and down over the runway's approach lights, I had closed the distance to the threshold, but now found myself too high. I rejected the brief thought of a go-around; I honestly felt I wouldn't be able to go through this a second time.

Reaching over with my left hand again, I pulled back on the throttle to cut power, but it wouldn't budge. Glancing down, I saw that my hand was on the mixture control. In my confusion, I hadn't sensed the ridges around its knob. I cursed myself; with 1,500 feet of the 4,800-foot runway gone, 20 feet over the runway, I had cut the power. Both hands on the wheel, my two left feet yawing the airplane wildly, I bounced the Cessna onto the pavement.

After doing S-turns down the runway for a while, I turned to my passenger and saw a huge grin on his face. He complimented me on the flight, especially the landing! I tried to explain to him what had taken place, but he wouldn't hear of it. He'd had a great time.

While tying down the Cessna I reflected on the many mistakes I had made. I should have heeded the warnings going off in my head as we taxied to the run-up area and not allowed the diminishing daylight to be a factor. I also should have aborted the takeoff when it felt like something was very wrong. And finally, I realized the lack of recent time in the Cessna contributed to my disorientation in the cockpit.

I found out a few days later that, although it's not against FAA regulations to fly from the right seat, it's not recommended without several hours of practice with a competent pilot in the left seat. Oh, really?

✈ ✈ ✈

*The pilot knew he was stretching his limits when he realized he was having trouble just starting the engine from the right side. "I continued on, thinking I would eventually shake the uneasy feeling." But he never did.*

*The next confession also involves a pilot who knew he was stretching both his limits and those of his Aero Commander 500. He, too, almost went too far.*

# Power Struggle

### by Tom Benoit

I'm a photographer, and this flight was to be a photographic safari from my home base, Gnoss Field, just north of San Francisco, to the parks of the West and Southwest. My friend Tim and I were planning to fly my 150-hp 1968 Aero Commander 100 over Yosemite, Death Valley, Zion National Park, Bryce Canyon and, of course, the Grand Canyon.

We got as far as Mammoth Lakes, California, elevation 7,200 feet, at the base of the Sierra Nevada. The wind was gusting to 25 knots, so the landing was rugged. We calculated our fuel and were pretty certain that we could climb out and land at an airport on the other side of the mountains to refuel. What we did not do, however, was calculate the density altitude.

The temperature was 75° F and the downdrafts were brutal. We added 17 gallons of fuel and calculated our approximate weight, did the run-up, leaned out the engine and started down the 7,400-foot runway. But at the halfway point the Commander was still not coming up to speed. Our airspeed never got above 53 knots, so we aborted.

As we taxied back we decided to try again. This time we'd keep the Commander on the ground as long as possible and try to gain airspeed. The airplane lifted off at 66 knots, but it didn't want to climb. For the next hour we flew over the airport trying desperately to climb, but we could only get 2,900 feet agl.

The air was turbulent. We decided to land, get a motel, and depart in the cooler morning air. The wind was now gusting to 30 knots, and the landing was rough. It was then that we checked the density altitude—it was 10,500 feet. Tim and I looked at each other in amazement; we realized that we were lucky to be alive.

The next morning we got out to the airport about 6:45 a.m.; the outside air temperature was 45° F. We were feeling pretty confident as we taxied down to the run-up pad. But when I pulled the brake handle, I felt something snap. We killed the engine, jumped out, and found that the hydraulic line on the right main gear had popped out of its fitting and was dripping hydraulic fluid on the taxiway. There were no mechanics at Mammoth Lakes, so we dug around the cockpit for a pair of channel-lock pliers and attempted to repair the line. Soon we felt as though it was repaired as good as we were going to get it. But we decided to avoid using the brakes at all costs. (This, of course, precluded aborting a takeoff should something go wrong.)

Once again, the Aero Commander struggled down the runway. When we finally pulled the airplane off, we were doing 70 knots, with a slight nose-high attitude. Everything seemed right. But at about 30 feet, the airplane began to mush and settle. A sickening feeling ran through me as I realized it was about to stall. We both knew it was on the edge but we didn't say a word to each other. Before the airplane could go into a full stall, we eased the nose over slightly. We were only about 15 to 20 feet off the ground, flying in ground effect at 70 knots. The Commander would not climb and would not gain any airspeed. I was petrified. If we hit even a breath of downdraft, we'd be finished.

We flew straight ahead on the high valley floor, until we approached a hill. Then we executed a shallow turn to the right, with the wingtip uncomfortably close to the ground, and continued to skim along the valley floor at 20 or 30 feet for about 10 minutes. Finally we reached 80 knots, eased the nose back, and began to climb.

After about 30 minutes we reached 10,800 feet and approached Mammoth Pass. The Commander was only doing about 70 knots and would neither accelerate nor climb. We hit another downdraft and were unable to hold altitude or speed, so we examined our options while flying around in circles. The problem was that the longer we flew around, the less fuel we had, and thus fewer options. We decided to go to Truckee, where we off-loaded approximately 100 pounds of gear and refueled. We took off with no problem and flew direct to Gnoss Airport. It's a flatland airport, and it sure was a welcome sight.

So remember that a little too much weight and inadequate horsepower can create a dangerous situation in the mountains.

✈ ✈ ✈

*Density altitude doesn't limit its influence to an airplane's wings. As the pilot of a Globe Swift found, it also impacts the pro-*

*pellers—airfoils themselves—and robs the engine of some of its power.*

# Not So Swift
## *by Robert Wick*

I had very little time in my middle-aged Globe Swift when I flew it to Grants/Milan Airport in New Mexico. Grants Airport is in a pass. The runway is fairly wide, and not difficult flying in the daytime. But it is high—6,520 feet above sea level. I had flown the Swift only once or twice around Albuquerque Airport, with its two-mile-long runways. While my log already showed close to a four-figure total, I was new to flying in the West.

The Swift was stock except for the engine. It had a 145-hp Continental, cannibalized from a wind-damaged Cessna 170, adding 20 hp over a factory Swift. Today, Swift owners often double the original power, but in those days 145 hp was pretty sporty. I'd taken the airplane to Grants for installation of leading-edge landing lights. With the work completed, I planned to pick up the airplane late one afternoon and fly back to Albuquerque. Business delayed me, and it was just dusk when I arrived at Grants Airport. No matter. I had plenty of night time, the visibility was reported as 85 miles, and there wasn't a cloud in the sky.

The run-up was routine, and last-minute checks showed everything to be okay. It had been a warm day and the evening temperature hadn't dropped much on the sun-baked asphalt. I carefully calculated the density altitude. It was close to 10,000 feet, but I was alone and not too heavy with fuel or baggage. With the extra horsepower and the Aeromatic prop, the Swift should handle this without much difficulty.

I lined up and shoved the throttle wide open. The runway lights of mile-long Grants began to slip past, slowly at first, and then faster and faster. The airspeed nudged the bottom of the green. Satisfied that I had minimum flying speed, I lifted off into the hot night air. The threshold lights fell beneath the nose as the hydraulic pump strained to put the gear in the wells.

The blackness to the left and right would have been foreboding, but almost dead ahead were the lights of downtown Grants. Beyond that were occasional headlights wending their way along U.S. 66, which, I thought, provided plenty of visual clues for easy night flight. But I was wrong.

In less time than it takes to tell, I was in big trouble. The Swift wouldn't climb at all. It staggered just out of ground effect, and no amount of coaxing would make it climb one more inch. The blackness on either side of the highway gave no clue about whether I had enough room or altitude to turn around and land. There was nothing to do but keep going. A quick cockpit check showed nothing amiss except slightly low rpm. But there was nothing I could do about it, since the Aeromatic prop decides for itself what rpm it will hold. I headed right down the middle of the main street.

The tops of the telephone poles seemed to brush the wingtips and the little Swift shuddered slightly, hanging on the very edge of a stall. Gingerly I tried to coax a slight climb but the buffet got worse. I lowered the nose just a hair. The Swift lost priceless feet. The airport dropped astern, but at least there was no danger of wires across the road between street lights. The only lights in that town are on the cars racing through the desert night.

Southeast along 66 from Grants, the desert falls away slowly until the road begins the long climb to Albuquerque's West Mesa. The Swift held its altitude while, ever so slowly, the highway dropped beneath the nose. After what seemed a lifetime, I had enough altitude to begin some careful experiments. A quick mag check showed nothing wrong. The throttle was already wide open, so that was no help. In view of the altitude, I carefully leaned the mixture slightly and found a dramatic improvement. The little Swift surged into a healthy climb. More careful leaning, and the Swift headed skyward with all the performance the handbook said it should have. With a sigh of relief, I climbed to 10,500 feet for the rest of the trip to Albuquerque.

What had I done wrong? That old bug, density altitude, had bitten—but not in aircraft performance. It had bitten in engine power. One of my sea-level habits was the culprit. For run-up and takeoff, I'd naturally pushed the mixture to full rich. While that's normal practice at most airports, it won't work high and hot. I was drowning the engine. Contrary to sea-level practice, it is important to lean for takeoffs at very high altitude. Fuel is used for engine cooling on takeoff, but a full-rich mixture is excessive above a density altitude of 5,000 feet.

Because the engine develops less power up high, it is unlikely to be damaged by leaning for high-altitude takeoff. Savvy mountain pilots taxi to the run-up area, add full throttle and lean slightly to peak power.

The Swift and I flew many happy hours together after that. I'd learned my lesson about high-altitude leaning, but where would I have been without that "big" 145-hp engine?

✈ ✈ ✈

*The pilot knew about density altitude. What he hadn't learned were the proper procedures for operating at high altitudes and hot temperatures. A short visit with a local instructor or a quick review of the pilot's operating handbook might have saved the pilot from having to learn his high/hot lesson the hard way.*

*While high-density altitude is a natural phenomenon that will rob an airplane of some of its performance potential, operating at high altitudes presents other problems. In the next confession, the pilot of a Luscombe 8A relates his experience in which a "gray curtain" brought on by hypoxia nearly signaled the end of his performance.*

# Smoke Signal

### by Scott Neil

I'd been waiting for an opportunity to shake the cobwebs out of my ancient Luscombe, a perky 8A of respectable vintage. Flying VFR in the Northwest is often limited to short hops of dubious good sense. It's more than common to find ourselves flitting here and there at the whim of the irritating Pacific storm track, seldom attaining any objective of real distance or arriving where we'd planned or hoped. And thus it is, when the sun finally bludgeons through the gray and the most insidious globs of moisture are fully 3,000 miles to sea . . . we fly.

I fueled at six a.m. The Luscombe was a trifle slow, dragging the weight of 24 gallons in an oversize tank. But the air was still and thick, and the airplane carried me faithfully skyward. I soared on to Port Angeles, 150 miles to the north; leaned at 8,000 feet, there was enough fuel for nearly five hours.

I leveled at 10,000 and headed for the Olympic Range. The wind was light, pushing crystal-clear air down from Canada. There was but the most dignified, gentle rocking of the wings through the great jagged crags, and then the airplane was flushed through the pass to Puget Sound. What a glorious day! I climbed to 12,000 and headed home, vowing to take half a turn around the apparition of Mount Rainier shimmering in the distance before settling back to the ground. Less than two hours later I was trimming to ascend the great volcano. Let's see how high it'll go.

At 13,500 I began to speculate on the effects of hypoxia. Not that any symptoms were present, or that I expected them, for air maintains a respectable viscosity even at such an elevation—I'd been there many times before. But I was a cautious pilot; I'd considered the potential, and that was evidence of good sense. I was calm and content. The old girl was flying herself, a rare event Luscombe pilots seldom encounter. I'd have a smoke. I fumbled for a new pack—darn it, I had to cut down— and I cursed myself quietly for such a lack of willpower. I'd consumed two packs of the stinking things already. I lit one and took a long drag. The noxious smoke rumbled down my windpipe, and I recall the beginnings of the anticipated nicotine "charge." Then there was a curious kind of explosion in both lungs, an unfamiliar sensation. I pondered it for a moment and then watched quite analytically as a gray curtain was drawn nearly over my eyes. It was the silliest thing.

In retrospect I believe I had some comprehension of the event. That bit of the consciousness that remains thinking to the end had reasoned that it could be little else: hypoxia. There was but one solution—go down. That scrap of remaining intellect then issued a command—exert a forward pressure on the stick—and the unthinking muscles scrambled to respond. Darn the gray. I shook my head to throw off whatever was covering my eyes. There was a notion of thickness—of the tongue, of the brain. There was a perceptible shaking of the airplane. The stick struggled mildly to jerk from my hand. I forced my eyes to see, and there was only blue. The Luscombe was fully stalled, twisting, trying to spin. Perhaps the easing of Gs allowed some blood to my starving brain, and for several instants I was granted some rational thought. I pushed the stick forward, and in a breathtaking tumble there was a flash of mountain and then the relative flat of a glacier settled into view, filling the windshield.

There are times when the mind accepts its apparent fate too easily. In those few seconds I could quite vividly envision the impact; an anticlimax in the deep, soft snow. There would be but a *wump* for no one to hear. Years would pass before they found the wreckage. The gas would have leaked away, and I was embarrassed that the world would think me that incompetent. Pilot error. He just ran out of gas. What injustice. What luck. What a ridiculous place to crash.

But there was some unknown independence in my muscles. While my conscious mind contemplated the event as a whole, some other force rose to action. The stick came back, slowly, properly. And at a hundred feet the Luscombe screamed down the mountain.

Some buffeting in light turbulence tended to revive my failing reflexes. I swooped into a large valley and circled below the rim, hoping, in a still irrational state, to find thicker air. I circled there, erratically diving and banking, sloppy on the stick. I strained to remember where I was and where I wanted to go. Down. That was enough for now. I snatched up a sectional and stared at the masses of lines and scribblings. I could remember marking a course from the peak to my home base, but I couldn't find it.

I lumbered on in this fashion for some time, demanding some solution, some sign of activity from my intellect, until at long last my head cleared. With every foot of altitude lost some new facet of my brain began to come on line, and by 6,000 feet there was a feeble semblance of confidence returning. I flew that last 50 miles white-faced and shaken, finally touching down in the old, familiar place. On the drive home the muscles of my stomach at last began to relax, and I was ill.

Shortly thereafter I kicked the habit, and every activity of life—flying included—has become a new adventure to experience.

✈ ✈ ✈

*Although he was aware of the dangers of hypoxia, the pilot had often flown at high altitudes and felt he could cope. He learned he couldn't. In the next confession, a VFR pilot finds himself in instrument conditions. He has two choices: He can turn around and land or he can continue to climb. Which would you choose?*

# Hoodwinked Over Oklahoma
## *by Marc Coan*

Most pilots would say that a low-time VFR pilot who takes off into instrument conditions, at night, with only a small amount of fuel on board and then doesn't try to land immediately has clouds where his brains ought to be. I agree, but I'd sure like to know how those clouds got inside my head one night.

The adventure began late one December evening, with a two-and-a-half-hour flight from Lawrence, Kansas, my home base, to Laverne, Oklahoma. I arrived at the unattended airport at 10:45 p.m. with about an hour's fuel left in the aging Skyhawk that I frequently rented. My plan was to drop off one of my two passengers at Laverne, fly 30 miles south to Gage, Oklahoma, refuel and immediately return home. My other passenger was a new pilot and would be returning with me. I had about 135 hours' experience and my friend had about 55. We were pressed for time because a cold front was forecast to move through the area later, lowering ceilings to 800 feet after midnight and to 400 or lower by morning.

The flight to Laverne was uneventful, with weather generally as forecast: 2,000-foot ceilings and visibility 5 to 10 miles. That is, except for one instance when I inadvertently flew through an isolated piece of scud about 25 minutes before we landed. I flew the entire trip under the hood to get a head start on my instrument rating, and my pilot friend served as safety pilot.

We saw our passenger off and then got back into the airplane for the quick flight to Gage. However, the airplane had other plans for us. The battery was apparently too weak to turn the engine over. This battery had given me trouble before and the wait-and-see trick had worked then. We waited an hour, until about midnight, and the engine fired instantly. A few minutes later we were rolling down the runway.

We had climbed only about 300 feet when I suddenly saw the landing light reflect a mass of white. The next thing I knew, we were surrounded by clouds. I reminded myself that the weather wasn't supposed to become IFR until much later; I thought I was in only a random piece of scud like the one I had encountered earlier. So, instead of turning the airplane around and landing at Laverne, I kept climbing, thinking I would break out in no more than a few seconds.

By the time I realized that this was no "little ol' piece of scud," it was too late. My chart showed three obstacles within a few miles of the airport, all at least 300 feet high. Lulled into a false sense of security because I had flown under the hood earlier in the evening, I decided I was better off flying in the clouds than down low in unfamiliar territory. Although I had no approach plates on board and no instrument approach experience, I didn't think it would be too difficult to shoot a makeshift approach into Gage using the VOR there.

Flying in the clouds turned out to be completely different from flying under the hood. Both my partner and I had to concentrate on the instruments for me to maintain control, and even then it was difficult. We pressed on and, although our course was erratic, we were soon heading outbound on a VOR radial that looked as if it would take us to Gage Airport. I descended to about 500 feet agl, but we never saw even so much as a rotating beacon.

Trying not to panic, I started a climb at maximum rate and headed east. My only plan of action was to try to break out into VFR on top so I wouldn't have to concentrate so hard on flying and could think about how to get out of this mess. At about 6,000 feet I saw the moon and immediately felt some relief. However, I never broke out. At 8,500, I decided I'd had enough and leveled off.

That's when I decided I couldn't do it by myself. I had to call for help. I got the proper center frequency for the region from an old en route chart I kept for training purposes, called "Mayday" and explained my predicament. The controller offered to give me vectors to Wichita, about an hour away.

"Uh, negative," I replied. "We don't have enough fuel for that. We need to find someplace closer."

He gave me a heading for Buffalo, Oklahoma, about 20 miles northeast of Laverne. I tuned in the NDB on Buffalo Airport and pointed the aircraft in the right direction.

Each minute that passed felt like an hour. The sweat from my palms made the control wheel feel slimy. Suddenly I saw the ADF needle swing toward the tail. The controller, who sounded as if he knew how to fly, told me to reduce power to begin a gradual descent while entering a 20-degree right bank.

I did as I was told. The altimeter began to unwind and the ADF needle spun slowly around as we circled the airport in a wide arc. When the altimeter read 5,000, the controller announced that he was losing radar contact with me because his coverage went down only to 3,000 feet agl in that region. I was glad he hadn't mentioned that earlier.

About a minute later we lost radio contact. I was in the clouds and descending into who knew what, and there was no longer anyone to offer me guidance or advice. Sometimes I caught myself rocketing downward at 1,000 fpm while in a 40-degree bank. Then I would overcorrect and find myself almost straight-and-level, or even in a shallow climb.

As the altimeter needle came closer and closer to the elevation of the airport, I slowed the descent rate. My partner kept his eyes focused outside the window, looking for obstructions, the ground or any sign of the airport.

We broke out at about 300 feet. My partner suddenly screamed that he saw a white flash to our right, and then we both saw a flash of green at about two o'clock. I pointed the nose right at the beacon.

I landed and taxied to the ramp. The air was much colder than it had been the last time we had been on the ground; the cold front had passed.

We found a place to sleep in an open shack that served as a pilot's lounge at the unattended airport. In the morning a local pilot happened to stop by, and he topped us off from his private fuel supply. The Skyhawk took 39 gallons, one more than the airplane supposedly had capacity for. I started working on my instrument rating as soon as I got home.

*Although the pilot quickly came up against his limitations, it was some time before he realized that he was in over his head. The pilot who relates the next experience wisely realizes the conditions are more than he wants to tackle alone. So he enlists his instructor to fly with him. The two learn an embarrassing lesson about personal limits.*

# Over the Hump
## by Richard Mazziotti

We didn't set out to wreck the Champ. In fact, we didn't even set out to fly that afternoon. But a few days earlier a nearby airport had reported a Cessna 152 missing. South of the airport, near where we kept the Champ, was a densely wooded ridge that I thought might swallow an airplane arriving after dark.

An air search of the ridge would be an interesting trip in the Champ, a wonderful airplane for low, slow flight. I put my wife through a thorough search-and-rescue briefing: "You look right and I'll look left." Then we took off.

We quickly learned a couple of things. The Third Armored Division could easily have hidden in that little ridge without anyone ever seeing it from the air.

In order to spot an airplane that has gone down in the woods, it has to cooperate by going in at a shallow angle and lopping off lots of branches, and it has to hang together enough to present an airplane-like outline of shiny metal.

Otherwise every old car body, of which there appear to be an enormous number in New Jersey, looks like a downed airplane. And if you do see a glint of metal through the trees, how do you get back to it and look it over again?

You don't. At least you don't when you're inexperienced and trying to follow someone else's hand signal (your inexperienced copilot's) pointing at a 50-acre swatch of timber and meaning, "It's over there, I think." Even when you wrack a slow and docile airplane like the Champ over onto its side, it takes many seconds and lots of acreage to turn around.

After about 20 minutes of fruitless labor we decided to gas up at the same airport that had lost the Cessna. The wind had started as a breeze earlier, but now the gusts were moving whole trees. It was blowing right down the runway, though.

Most of the airport loungers found something to say about our landing, but I thought we'd settled down fine after the third bounce.

While refueling, we met our instructor, who also felt compelled to comment on my vigorous and various landings. At that point I made decision number one, a decision I've always been proud of: because of the wind, we would leave the airplane and hitch a ride back to the 1,600-foot grass field where we based the Champ.

Then I made decision number two. What the heck, I thought. Why not fly the airplane back with my instructor and get some crosswind landing dual? My wife could drive his car there to meet us.

The home field was a sloping, wide grass strip with nothing at the upper end; and a barn, trees, house, outhouse, garage, picnic table and airplanes at the lower end. In the middle was a hump of earth of possible interest to archaeologists but of no earthly use to landing airplanes, except to instill mirth and snickers in the loungers that seem to hang around every airport, no matter how small.

My instructor and I arrived over the grass strip as planned. The waving trees and wheat seemed to indicate that although the wind was just about perpendicular to the field, some of the gusts showed that landing downhill would be more into the wind than uphill.

I flew downwind, base and final with sweaty hands. I added a few extra knots to my approach speed and made firm decision number three that we would just "take a look and probably go around."

On final I made the decision to land. We touched down on the first third of the field, and we even stayed down for a while. Then we hit the hump in the middle, and suddenly we were 10 feet off the ground.

I started to make the fatal mistake of pushing the nose down to get the airplane back on the ground when my instructor slammed the stick back. And we hung there, and hung there, heading downfield.

By the time we flopped back to earth we had used up a significant portion of the runway, and my instructor yelled that we must groundloop to the right to avoid hitting the airplanes and the barn that we were rapidly approaching.

Right stick, right rudder, right brake—all were extended to the stops. In front of us my wife, the airport's owner and several others who had come out to see the show scattered wildly.

The Champ started yawing to the right, but we continued skittering on the grass straight for the barn. At the last second the airplane seemed to sort of shake itself, stopped moving forward and turned 180 degrees to the right, burying the left wing and tail in the pine trees that lined the turf runway.

We sat there quietly and watched the prop go around. We had not even shut off the engine.

At first I was just stunned. Then I hoped that we had no damage. I looked out the left window at the bent strut and was stunned some more.

The left wing bow, spar and strut were all damaged, but replaceable or repairable. The left elevator was crunched, and the Champ's tail feathers were slightly askew.

Looking back I see I made too many decisions that were based on wishes instead of facts. The extra speed I carried on final (half the gust strength plus pucker factor) was too much, and I never should have landed downhill.

But I also learned that an intentional groundloop can be a lifesaver. It is far better to hit the trees sideways than the barn head-on.

# 10

# Navigate, Communicate, and Aviate. . .

As long as all the gauges and electronics are performing their functions, pilots can know which end is up and where they are, and they can tell other people. Many problems that develop result when something on the panel quits functioning and leaves us without vital information. Or it malfunctions and we don't realize it and fly along, believing the information we're being given.

Cockpits have become more and more sophisticated over the years. More instruments and electronic gadgets make the pilot's workload easier—providing they're operating properly. Of course the more complex the panel, the more opportunity for something to go wrong.

In the first account, the pilot was faced with navigation radios that received some navaids, but not others. Finding the right combination of frequencies that the radios could receive and the approaches available to complete the IFR flight put the experienced pilot behind the eight ball.

## Radio Daze
### by Joseph Burnside

The first hint of trouble came at the initial hand-off. After dialing in the appropriate frequency in the number-one com, all I could hear was faint music leaking through static. The number-one nav, the one with the glideslope, had in fact never really locked onto the VOR some 20 miles ahead.

No problem, I said to myself as I switched to the number-two nav. I have two radios: Even if the number-one never comes on line, the number-two is fine, and the ceiling is well above the localizer-only minimums for the approach. Although a glance outside showed nothing but the inside of the stratus I'd been in since liftoff, I knew it would still be okay.

I had just departed an airport in southwestern Pennsylvania after dropping off a friend for his golfing weekend. The weather at my destination and the airplane's base, Manassas, Virginia, was forecast to be the same as when I had left two and a half hours before: 1,200 feet overcast, five miles in fog with light drizzle layered up to the flight levels. I'd flown through this stuff to get here and was not anticipating anything more on the way home than an hour or so of playing "pinball" with the needles and gauges while changing frequencies. Little did I know.

After the number-one radio started acting up, I used the number-two radio for the next 45 minutes, occasionally switching the number-one on and off, hoping the problem was a momentary glitch. Soon I was on a 30-mile vector to intercept the localizer at Manassas. By this time, I had tuned the number-one to the localizer frequency, had the approach plate out and was reacquainting myself with a very familiar procedure.

The CDI on the number-one was still waving at me, though, and as I turned up the audio in an attempt to identify the station, I was greeted with Bob Seger's "Hollywood Nights"—*not* a good sign. Although the Skyhawk I had borrowed needed some money invested in its engine and avionics, I had thought I was familiar enough with its idiosyncrasies. I was also quite proficient on instruments. Plus, there were plenty of airports with approaches along the way. If one radio went out, I had reasoned, I always had two, right?

Well, I thought, I'm still a good way out; maybe it's not receiving a strong enough signal. The number-two was still good; the ammeter showed nothing amiss, and the weather was holding. I decided that I'd continue to the final approach fix (FAF), and if the number-one was still off-line, I'd just switch to the number-two and the localizer-only minimums.

That's precisely what I did. After Dulles Approach dropped me down and cleared me for the ILS, I read back the clearance and added that this would be a localizer-only approach. As I looked down at the number-two CDI, I noticed with satisfaction that it was centered. But then I saw the red OFF flag. *Definitely* a problem.

"Cessna 85 Hotel, I show you a bit right of the localizer, turn 20 degrees left," the controller suggested.

"Uh, approach, we don't seem to be able to receive the localizer. Are you monitoring it okay?"

"We'll check on that 85 Hotel. It's monitored locally."

"Roger, ah, 85 Hotel, missed approach," I responded as I added power for the go-around and began to climb back to 3,000 feet. Settling on a vector to reintercept the localizer, I started twisting knobs in earnest. As I again sought to identify the localizer, a report on the area's rush-hour traffic leaked through on the number-one nav while the second one remained silent.

"Ah, Dulles, 85 Hotel is unable to receive the localizer at Manassas on either radio. We are receiving Armel just fine; requesting the VOR-B into Manassas." Armel is the closest Vortac to Manassas and sits just west of the main runways at Dulles, some 12 miles north of Manassas. It provides a step-down and final fix for the approach I had just missed and course guidance for the seldom-used VOR approach.

"Eight-Five Hotel, I can give you the VOR but will have to vector you up toward Leesburg," responded the controller. I thought about that one for a moment. I had planned the flight to have about an hour of fuel remaining when parked at the tie-down. That made me legal to file Dulles as the alternate and have the requisite 45 minutes remaining. The plan was holding: I should have about an hour of fuel on board. But the vector to Leesburg and back would take 20 minutes or so, resulting in that much less fuel if I missed, a real possibility. Not what I had in mind. I glanced at the fuel gauges. They were bouncing about a needle's width above the big red "E."

"Negative, Dulles. How about an immediate ILS to Dulles?" I didn't want to be flying around wondering where I would be landing any longer than possible.

"Okay, plan vectors to the ILS 1 Right at Dulles." As I received instructions to turn and descend I found the plate and dialed the frequency into the number-one nav. It waved. I dialed it into the number-two. Nothing. "Eight-Five Hotel, this is now radar vectors for the ILS/DME 1 Left at Dulles."

"Ah, Dulles, 85 Hotel is negative DME," I responded, slow to catch on to what my nav radios were telling me.

"Okay, plan the right side."

Still nothing. "Dulles, 85 Hotel is unable any localizer frequency at Manassas or Dulles. I don't know what the problem is. Requesting the airport surveillance radar [ASR] into Dulles."

"Eight-Five Hotel, the ASR at Dulles has been decommissioned." Wonderful, I thought. I had flown out of Andrews AFB, on the other side of Washington, D.C., a few years earlier. They had ASR and PAR [precision approach radar] but also lots of questions and paperwork. I'd have to use the "E" word to get in. More paperwork.

"Eight-Five Hotel, I can give you the VOR into Manassas right now. I do have traffic inbound for the ILS, but I can slow him down."

Bless you, kind sir. "Yeah, we'll take it," I responded as quickly as I could while trying to maintain my *Right Stuff* radio voice. I quickly dialed Armel into the number-two nav, then turned and descended as requested. I was almost on top of the final approach fix for the approach. After rotating the OBS to the correct radial I simultaneously noticed a two-dot deflection to the left, leveled at the proper altitude, banked to the left and was cleared for the approach from over the FAF. After noting the time, I started a descent, broke out some 600 feet above the minimum descent altitude and aimed the bird where I thought the airport was. Sure enough, after about 90 seconds, I spotted the airport beacon.

"Dulles, 85 Hotel has the airport, canceling IFR, very much appreciate the assist," I sputtered. After an abbreviated pattern, I was able to slide the mains onto the runway and make the first turnoff. The whole affair hadn't lasted 10 minutes.

As I taxied in, I thought about what had just transpired. On departing earlier in the day, both radios had checked out fine. En route to Pennsylvania, the airplane performed perfectly. Both VORs matched, and there had been no indication whatsoever that the radios would act up later.

Pushing the Skyhawk into its tie-down spot, I noticed water dripping from the belly. I had flown through a lot of rain, but hadn't considered it a problem. Now I can only guess that water had leaked in, fouling the coupling of the two nav radios into the single "cat's whisker" VOR antenna on the tail. Why that might have affected the localizer frequencies and not Armel's, I don't know. Also, it doesn't explain the failure of the number-one com.

I do know that I came close to running out of options that afternoon. Proficiency in the IFR system, strong familiarity with both the airplane and the area coupled with plenty of experience flying single-pilot IFR helped. More fuel would have made the situation less immediate. The bottom line, however, was that the rate and condition of radios dictated a much more conservative outlook on operating that airplane in less-than-good VFR. I also know that what began as a marginal trip had crossed over the line from possible to potentially catastrophic.

Contemplating all of this, I drove home more slowly than usual, vowing not to get as close to that line again.

✈ ✈ ✈

*The pilot's bacon was saved by the number-two radio still being able to pick up the VOR for the approach. Silence is golden but not when you're flying on instruments. As good as the lost-communications procedures are, having a backup handheld transceiver has to be very comforting.*

*In the following incident, water again causes a loss of communications and navigation equipment. This time, the Air Force instructor pilot earned his month's pay and demonstrated the value of dead reckoning.*

# Dead Reckoning in the Clouds
## by James R. Walters

It promised to be a fun-filled training flight. My student pilot and I had a USAFT-38 aircraft and fuel credit card at our disposal. The lesson objective: to teach the student pilot how to navigate within the high- and low-altitude IFR structure.

I momentarily recalled the days way back when I learned navigation as a student pilot flying Cessnas. My instructor-father taught me that the primary means of navigation was dead reckoning—*everything* else, he said, was either a refinement or a backup.

We blasted out of our west Texas base during early morning. Towering cumulus were building out of low stratus, and the freezing level was reported at 11,000 feet. Our plan was to first fly high and fast to Columbus, Mississippi, where we'd make a pit stop, then fly a victor airway leg to our final destination, Tyndall Air Force Base in Panama City, Florida. Tyndall was 400 overcast, with one mile in drizzle, conditions that offered my student the great opportunity to gain some actual IFR experience—something he hadn't seen much of in west Texas.

The first leg was uneventful; the second leg proved to be more interesting. Traveling southeast out of Columbus, Mississippi, we passed over Montgomery, Alabama. From 9,000 feet we could look down through a broken layer with bases at

3,000 feet and see Dannelly Field. Then we turned south, right into solid soup, where we would remain all the way to Tyndall.

Steady rain beat on the canopy. Flying in such weather would give my student good IFR experience, I thought. Contentedly dry in the cockpit, we flew on with, like most pilots, great confidence in the radios housed in our nose-cone compartment. One com and one nav (tacan), an ILS, a transponder, and a single-channel backup transceiver tuned to 243.0 (the UHF frequency known in the military as "guard') gave us a sense of security, albeit a false one.

The rain continued. Then, without warning, the NAV and DME flags suddenly flipped to OFF. The HSI simply died.

I radioed our plight to Tyndall. "Approach, Roman Five-Five. We've lost our tacan and directional gyro; cancel our tacan approach. Request vectors to a no-gyro precision full stop." At this point I considered executing a no-gyro approach to be just more good experience for my student—that is, until we were 19 miles out and the voice of the final controller turned into the sound of bacon frying.

I switched to the backup channel. "Tyndall GCA, Roman Five-Five. We've lost communications. Request emergency GCA on guard channel."

"Roger Roman Five-Five, GCA on guard. Turn left heading 340; one-three miles from touchdown, slightly right of. . . *scratch. . .pop. . .sizzle. . .*" More bacon frying.

The weather conditions might have been just right for a student seeking actual IFR experience, but they were not quite right for our navigation "backup" equipment. We were unaware that a hole in our fiberglass nose cone had been expanding from the size of a pinhole to the diameter of a quarter during the rainy flight. Heavy water was seeping into the radio compartment, and that water was shorting out and damaging our com and nav equipment—the reason for the sounds of bacon frying and the gyro instrument failure.

In the silence, I became suddenly aware that this was one of those decision-making moments during which a pilot earns his whole month's pay. I quickly reviewed our options. There were four ways out of this mess. We could press on in the blind, and let down to circling minimums, knowing that Category E provides obstruction clearance of 300 feet for 5.2 nm; but the problem here was that Tyndall circling minimums were 300 feet up in the clouds.

The second option was to turn out over the water, let down to VFR, then run the scud back to the beach. That might work on lots of days—but not on a day with only 400 feet and maybe a mile visibility. As an extra added attraction, there were also lots of high yard-arms in the Panama City area.

Option three: Use the ejection seats and let the airplane crash. With 50 minutes of fuel left and the thought of facing the post-ejection paperwork required by the military, this was not a particularly promising alternative. I chose option number four and headed for the nearest VFR airport, Dannelly Field at Montgomery, 140 nm back to the northwest. The problem now was just how accurately I would be able to find Montgomery using only a magnetic compass and a clock.

Jamming the throttles to 100-percent power while pulling up into a climbing left turn, I got on the intercom and asked the student to set the squawk to 7700, climb to freezing level and level off away from IFR altitudes. He was to hold our last vector heading, 340, until I computed a better one.

"Anybody on guard, this is Roman Five-Five, transmitting in the blind," I radioed. "We are departing Tyndall to the north, without com or nav, expecting to arrive at Montgomery in 30 minutes, emergency fuel." I listened for a response. I heard only the sound of bacon frying.

I had my student continue flying the airplane, not so much to give him more experience (he was already learning by the minute) as to keep me free to calculate, navigate and think. After 17 minutes I announced, "Pull the engines back to idle and descend to the MEA. Montgomery should be dead ahead."

We broke out at 3,000 feet, with 15 minutes of fuel remaining, and caught the sweet sight of a divided highway running north/south just off our left wing. There's just one of those in this part of Alabama, I thought, and it runs right up to Montgomery—which was beginning to emerge from the haze ahead. A green light flashed from the tower as we touched down.

During postflight debriefing I reminded my student, "Now the real lesson here is that the primary means of navigation is always dead reckoning. Everything else is either a refinement or a backup." But I knew he had already learned this lesson well.

✈ ✈ ✈

*When his panel went south, the T-38 instructor pilot was able to "reckon" where he was by calculating time, speed, and distance.*

*Surprisingly accurate, reverting to the basics has often bailed out pilots when their state-of-the-art instruments go on strike.*

*Partial panel is a skill that needs to be polished and practiced frequently. In the next incident, a pilot practices his partial-panel techniques and is almost immediately rewarded by an opportunity to put to good use his so-recently honed talent.*

# Slave Meter Savior

## by Bernard Leewood

My wife, Betty, and I were planning a flight from our home in Chapel Hill, North Carolina, to visit relatives in Shreveport, Louisiana. The forecast for Chapel Hill for the morning of our flight was for low ceilings and light rainshowers. However, we arrived at the airport to find VFR with breaks in the overcast. We were grateful for the dry skies that greeted us; though we love the Mooney, loading baggage in the rain tends to dampen one's enthusiasm for low-wing aircraft.

We got our clearance from Raleigh-Durham Clearance Delivery and were off at 9:30 a.m. for Huntsville, Alabama, our refueling stop. The club-owned Mooney 201 was equipped with autopilot, RNAV, HSI, a backup vacuum system and, for the icing on the cake, loran. Except for CAT II capabilities, we were as well equipped as I had been in the DC-10s I flew for an airline until my retirement in 1985.

We flew the first leg at 8,000 feet and went on the gauges about 50 miles out. With a few vectors to keep us out of Charlotte airspace (which also put us 20 miles north of our planned route), we settled down to a smooth, relaxing flight. Just before reaching Asheville, North Carolina, I began to hear aircraft reporting moderate to severe turbulence and requesting detours to the south. My Stormscope showed nothing, so I declined any detours. We broke out on top near Asheville and soon started picking up light to moderate turbulence. We tightened our seat belts and shoulder harnesses and rode through 20 minutes of light to moderate chop. The remainder of the flight to Huntsville was smooth and uneventful.

The weather at Huntsville was as advertised, 400 and two, and all the goodies worked for the coupled ILS approach to Runway 36L. We stayed on the ground just long enough to stretch our legs, top off the tanks, get the latest weather (Shreveport clear and 20, light tailwinds all the way) and pick up our new Icom transceiver in the luxury of the pilot's lounge.

All was right with the world, and to boot, we'd have a tailwind heading west.

We took off on 36R, got the gear up, established a cruise-climb, switched over to departure control, and were given a southwest vector "direct GLH when able" with a clearance to 6,000 feet, my requested altitude. Going through 400 feet, I was back on the gauges. After about 30 minutes on instruments I began to feel a bit guilty, not practicing partial-panel in the Mooney as I had admonished many of my students to do. I pulled a couple of rubber discs out of my flight bag and covered up my HSI and attitude indicator, then knocked off the autopilot, determined to practice what I preach. Since I was in actual instrument conditions, I didn't have to go through the discomfort of wearing a hood. After 10 minutes of rather tedious and tiring practice, I was ready to go back to the easy life of the autopilot, confident of my partial-panel skills.

Keeping my scan on needle, ball and airspeed, I removed the rubber discs. As I took the disc off the attitude indicator it indicated nose low a few degrees. I made a mental note to write up the misalignment when I returned to Chapel Hill at the end of our trip. Just a few seconds later, with my scan still needle, ball and airspeed, the AI showed a diving turn to the right, though my scan told me "not so!" It seemed improbable that one minute after practicing partial-panel I would actually lose my vacuum, but one glance at the annunciator panel and the flashing hi-low vacuum light (indicating a low vacuum) made me a believer.

I pulled the knob to activate the back-up vacuum system, but it was somewhat difficult to move (I later found out that the cable was kinked in its housing and immovable). I tried several times to engage the backup system, but the AI still showed a diving right turn and the annunciator light kept right on flashing. "Just great!" I thought, "my emergency backup is inoperable, now it's needle, ball and airspeed for real—no more fun and games."

Looking at the turn-and-bank, I saw that something just wasn't right. The turn needle didn't move. It was frozen dead center, but the HSI was indicating a turn. Of course, I wasn't sure of this either since DGs are normally vacuum-driven. There was only one explanation for this—the turn needle must be part of the failed vacuum system. So what about the DG portion of the HSI; is it electric or vacuum? The letters DC were legible on the casing of the HSI, but given the panic I began to feel, I had to make sure, and quickly. To check the accuracy of the indications, I looked far right on the instrument panel to the slave meter. Since the HSI at that second wasn't moving, the slave meter should be centered, and it was! A moment later I noticed a turn on the HSI. Looking back to the slave meter it had moved one index line to the right. I applied left aileron until the slave meter was centered. After a while, I found that by keeping my feet off the rudder pedals and applying light aileron pressure, I was able to keep the airplane from turning by keeping the slave meter centered. Since the airplane was trimmed for level flight and I was in relatively smooth air, with a light touch on the control wheel, taking the rate of climb into my scan, altitude control within 50 or 100 feet was no problem.

When I had the aircraft squared away, I notified ATC of my predicament. When asked my intentions, I told them I understood weather was severe clear at Shreveport and I was going to maintain my present course direct. I was cleared RNAV direct and was to inform ATC when I could maintain VFR. I broke out in about 25 minutes around Greenville, Mississippi, and was almost hesitant to tell them; I was enjoying this direct-to-destination clearance. However, after a few minutes of soul-searching I informed ATC I could maintain VFR from this point on. Much to my delight, they cleared my present position direct to Shreveport.

During the return trip to Chapel Hill, after the installation of a new vacuum pump and the repair of the backup vacuum cable, I practiced some turns with the slave meter. I found that, depending on the direction of flight, you either correct toward the needle or away from it. I determined that one index line was about 10 degrees of bank and two, approximately 15 degrees of bank. I would imagine this varies with different HSIs or slaved compasses, but that's how it worked on N3874H.

Upon reflection, I can't help but wonder how much luck or what part Providence played in my experiences. Had the vacuum failure happened 20 minutes earlier, it would have been on the instrument approach to HSV, with a different outcome, I'm sure. What are the odds of having a systems failure within minutes of practicing that procedure?

I have never read anything that encourages the use of the slave meter under such circumstances. However, if necessity is the mother of invention, then surely self-preservation is the father of ingenuity in the cockpit.

✈ ✈ ✈

*The Mooney 201 pilot, having practiced partial panel, was able to fly straight and level by keeping the slave needle centered. But just being sure the needles are centered won't always keep you out of trouble. Centered needles nearly led an instrument instructor, distracted by his video camera and lulled by his near-perfect student, astray.*

# Beware of Centered Needles
## by Jim Giordano

It was one of those fall afternoons in Michigan: 4,000 scattered, bright blue above and visibility unrestricted. My student and I were about 30 minutes into the instrument training mission, and the cockpit had settled into the comfortable process of practicing instrument approaches.

We were about 10 minutes out for the downwind for the ILS 27 approach into Flint. The student was one of those natural flier types. He had adapted to instrument flying well and had recently begun keeping the glideslope and localizer needles of the Cessna 172 centered the entire ride down the approach.

I had watched this instrument candidate progress to a confident and smooth flier, and he had gained my confidence, my complete confidence. We were at the stage where all the procedures I had gone over in my mind were acted on by him. For the last four flights, I'd had relatively little to add in the way of instruction. We were putting the final touches on his instrument training. As sometimes happens with overconfidence in a good student, my mind began to wander.

Because of the advanced stages of flight training for this pilot, I had brought along my new hand-held video camera and had figured out a way to record the voice portion through the intercom. My hopes were to create an instructional tool by video-taping the flight instruments, air traffic control communication, our responses, and the cockpit conversation.

With a competent student in the left seat and this new video gadget in my hand, we were cleared to intercept the final approach course. As the localizer needle began to center, approach control abruptly turned us north for spacing. Secretly I was pleased. This would allow more time to practice holding the camera, checking for traffic and monitoring the student.

It was the kind of day that brings out many IFR training missions in Michigan. The frequency was alive with instructions and responses. My thoughts began to wander and I imagined that I might even use this tape for the next instrument ground school. After some time, we were turned south for another intercept. I brought the camera to my shoulder and began recording.

Instantly I realized this would not be as easy as I first thought. Keeping the instruments in the camera's viewfinder while looking for traffic would be next to impossible. But I didn't worry; we were in radar contact and the student under the hood was a competent pilot.

"Seventy-Five Bravo Gulf, turn light to heading of two-nine-zero, maintain 2,700 until established, cleared for ILS," approach control spoke. The student responded with a flawless read-back. This was going to be great footage.

I was looking through the view-finder, focused on the primary flight instruments, the localizer and the glideslope. Perfect. The needles were now centered. I'm not sure what made me look up. Perhaps it was luck, perhaps it was instructor instinct; however, I had the feeling that something just wasn't right.

My eyes went immediately to the panel. The flight instruments looked normal. The altimeter said we were getting very near to decision height (200 feet agl). Something came into my peripheral vision. A tower slipped by the window, level with the cockpit. "Missed approach," I said loudly, dropping the camera in the back seat and applying full power.

Now my full attention was where it should have been. The airport was dead ahead; however, we were still two miles out. As we began the climb out I tried to figure out what went wrong. How did we get so low with the glideslope and localizer needles perfectly centered?

It took only one scan to realize what had happened: The glideslope flag, centered and looking alarmingly normal, said OFF. Due to the vector for spacing we'd been flown on a long downwind. Sometime during the localizer intercept the glideslope had failed. Having thought the intercept was good, we'd begun a 500-fpm descent. Maintaining the 500-fpm descent, the student thought he was flying the glideslope. Through my viewfinder, the needles appeared centered just as they had been for many other approaches this student had made.

I learned several lessons that day. First, no matter how experienced you are, never allow your full attention to become distracted from your primary mission—flying the airplane. Second, no matter how well a student is doing, as an instructor you are the pilot in command. Beware of the "halo" effect which lulls you into believing a student, even an experienced one, doesn't need your assistance. Third, always check the flags. Many glideslope indicators park in the center when flagged, giving the appearance of a perfect approach. Finally, make sure the safety pilot is always looking out the window during simulated IFR flight.

By the way, I have some nice footage of an approach that almost turned out to be very instructional indeed.

*The Cessna instructor was inattentive due to his experimentation with the videocam. Distraction can often lead to disaster. The pilot of a Rockwell Commander was distracted by a heated argument with his wife, and let it get in the way of cool professionalism in the cockpit. There was nothing wrong with his instruments. He just ignored three warning flags waving to get his attention.*

# It's My Wife's Fault
### by Bob Utterback

It happened on an IFR night departure from Fresno, California, and I was more frightened than I had ever been in my life.

After four years as a regional sales manager with Rockwell Commander, I moved to Oakland, California, in 1978, to work for a dealer. My friend Bud, a corporate pilot in Fresno, called one day to say he had a prospect for a used Turbo Commander.

We had a 680T in inventory with fresh paint and a new interior that would show well. The airplane was equipped with a Collins flight director, dual King Gold Crown navcoms, King DME and RCA radar. The former corporate owners had spared no expense when they originally bought N691W. (For the benefit of readers who have not flown with one, a flight director combines the information that is usually acquired from a scan of several different instruments. Artificial horizon, heading and navigation inputs are all presented in one display, rather than separately, and the flight director is usually placed in the center of the pilot's scan to give him a quick picture of the flight situation.)

Two days following Bud's call, I left for Fresno with my wife,

Pat. On arrival, Pat joined Bud's wife while he and I planned our sales presentation. We had planned to pick up the prospective buyer in Palm Springs and return to Fresno. But as it turned out, the gentleman wanted to first take his new wife, their two-year-old son and his sister-in-law to San Diego for a late lunch. Although I would have preferred to return directly to Fresno to rejoin my wife, I realized my job was to sell the airplane.

By the time the party returned to Fresno, it was after dark and an overcast had formed. I was angry with the gentleman for having caused the delay and rather guilty about leaving Pat behind in Fresno all day. After filing an IFR flight plan back to Oakland, Pat and I walked to the Commander, exchanging heated words over my delay in getting back. These hostile feelings set the stage for what was to follow.

I took down my clearance from the ground controller and slammed the old Turbo Commander out to the end of the runway, trying to vent my frustrations on the airplane. The tower cleared N691W for takeoff. I wondered why the heading on the Collins flight director did not agree with the runway heading, but I was too upset to consider the matter further and fire-walled the throttles.

As we departed Runway 18 and got the gear up, I started a gentle turn to the right to intercept the northerly radial that my clearance required. The bottom of the cloud layer was around 1,500 feet with good VFR underneath, and the city lights out below commanded more of my attention than the instrument panel, which I should have been thoroughly scanning.

As we passed through 2,500 feet in the clouds, I suddenly became aware that the airspeed was increasing in spite of the back pressure I was holding on the wheel. The usual climb speed of 130 knots had now passed 160 knots. Then the altimeter no longer registered a positive climb. I sat there with my mind spinning, trying to correlate these two facts with the information from the flight director, when I caught a glimpse of some lights through the overcast.

Remembering that the only lights around during climb-out were on the ground caused the hair on the back of my neck to stand on end. The lights were in the upper right quadrant of the windshield! How could this be? By this time, I was completely disoriented. I had no idea of my heading or of how close we were to the ground. Panic was imminent.

More lights appeared, and I could see we were in a nose-down 60-degree bank to the right. A glance showed the airspeed going through 200 knots—and the altimeter unwinding madly. With visions of encountering an obscured hill in the night, I stepped hard on the left rudder, snapped the wheel to the left and back. As we were pushed down in our seats by the pull-up, I yanked the throttles back to bleed off some airspeed. Then, as the real horizon slowly righted itself and I started to breathe again, I glanced at nonpilot Pat. All I could see were two wide, staring eyes.

With the immediate threat over, I started to analyze how all this had happened. Then it dawned on me: "You idiot, you forgot to switch on the inverter to supply ac power to the flight director." I had been following a dead instrument—despite the fact that three warning flags were up—and was literally flying by the seat of my pants in the clouds. Worse than that, I had violated one of the cardinal rules of instrument flying: Scan the instruments. . . scan ALL of the instruments, ALL of the time.

That flight brought home a few never-to-be-forgotten facts: Flying an airplane demands a pilot's full, never partial, concentration. If anything is likely to detract from that total concentration—whether it's fatigue, a recent argument, or whatever—*don't fly*. Also always use a checklist—if I had been meticulously going through the items on the list and not been so angry with the airplane, it's likely the error would have been discovered.

When I flipped the inverter switch, the flight director quivered to life. The three warning flags—which had not registered in my brain because of my agitated state—disappeared from view, and the horizon, heading and nav information were again displayed, this time accurately. The remainder of the flight to Oakland was made safely—and with a lot of silent prayer.

✈ ✈ ✈

*The Rockwell Commander pilot let things distract him, so he didn't realize the instruments weren't working the way they should. The pilot of a Piper Malibu in the next account found himself in a situation where everything was working the way it was supposed to. It just wasn't enough.*

# Slow Going in White Weather
## *by David Mangone*

The clock radio went on at 5:30 in the morning, waking me for the flight I had planned from Santa Monica to Sacramento. On the way to the shower I glanced out the window; it looked like a miserable December day. It was overcast and misty, and a foggy haze hung over the water in the marina.

For a seven a.m. departure the briefer reported conditions to be 600 overcast. Climbing out of the basin I could expect light to moderate turbulence, and I would be on top at about 10,000 feet. En route weather was to be clear at my cruising altitude of 16,000 feet, but there was a brisk 50-knot headwind. For landing at Sacramento, winds were gusting to 30 knots with reported wind shear, and the airport was IFR, but well above minimums. So far, although it didn't sound like the most enjoyable flight I could take, there was nothing in the report that would preclude me from flying my Piper Malibu on the round trip. But just to be sure, I specifically asked two more questions of the briefer: What's the freezing level? And what about icing? He said, "Freezing level is 7,000. There is no icing forecast, and none is reported." (The Malibu was certified and fully equipped for deicing, but I like to be forewarned.)

I turn on pitot heat and stall warning heat as soon as I even see a cloud layer (this time they went on in the run-up area). Takeoff was normal. True to the report, as soon as the wheels came up, I was in the clouds. I made my turn, set the heading bug, dialed in and verified the 323-degree radial from LAX, set power and trimmed the airplane for cruise-climb, and switched on the autopilot.

I glanced out the window; the view hadn't changed from a solid blanket of white since takeoff. But wait . . . what's that frosty stuff starting to collect on the wings? Maybe that didn't really count as ice, since the briefings I'd gotten were quite specific on that topic. Still, I added window and prop heat and noted the little jumps in the ammeter that would tend to verify they were indeed on. And I made a mental note to keep an eye on the wings as I turned my scan back to the instruments. It takes a quarter to a half inch of ice on the wings before the de-ice boots can be effective; any attempt to exercise them before that point can be counterproductive, since it may create a cavity between the boots and the ice, which could render the boots ineffective from that point on. But I felt I was a long way from that; this was just frost.

My climb continued out of 6,000 feet for 16,000 feet as I again scanned all the instruments. I had been climbing at 1,000 fpm, but now I was down to about 500 fpm with an indicated airspeed of 120 knots. That didn't seem right. I checked the altimeter; I was just below 8,000 feet. I remembered to check the wings again. As alert as I thought I was, I suddenly became *very* alert. It couldn't have been more than a minute, maybe two, since I had last checked outside, and that little frost now looked like solid ice and was definitely approaching the half-inch limit.

I activated the deice boots and brought the engine up to full power to add some airspeed. The ice stayed put, the airspeed stayed put, and the rate of climb was slowly falling to zero. The light chop I had been experiencing suddenly changed in character; my senses strained to identify it. I've practiced lots of stalls in this airplane and this didn't feel quite like a stall. But whatever it was, the airplane was not happy.

Looking up, I could see patches of blue sky on top, but I couldn't get to it. I turned off the autopilot and took the controls myself. This must be some type of buffet preceding a stall, I thought, but how could I be stalling at 120 knots, nearly twice the stall speed of the Malibu? I lowered the nose to keep the airplane flying, heading straight for a mountain I knew was dead-ahead, with the engine temperature over redline. I kept thinking to myself, how did this happen so quickly?

I called Los Angeles Center, telling them of my problems and asking for lower terrain. They directed me to turn 90 degrees to the right and maintain 8,000, asking if I wanted to try for Lancaster (Fox Field, which is in the Mojave Desert and would not require crossing any more mountains). As I dipped my right wing, the airplane wanted to fall out of the sky; now I was fairly certain that this was a wing-icing problem. I put all my concentration on flying the airplane, forcing myself to make a much more shallow turn than I wanted and, despite my desire for altitude, lowering the nose for a little more airspeed.

I was still transmitting to center, telling them that Lancaster sounded fine but that there was no way I could maintain 8,000; the best I could do was a slow descent.

I finally broke out of the clouds at 4,000 feet over the desert, with the airport right in front of me about five miles away at 2,400 feet. I must have looked like a ghost ship shrouded in ice.

Nevertheless, when I contacted the Fox tower—which supposedly had been made aware of my situation—the controller put me number two following a 152 on right base.

Now came the mistake that could have ended my flight prematurely. I didn't think of it as a mistake until I reviewed my performance later, but it was a definite mistake, and it was born out of politeness exhibited at the wrong time. I thought, the 152 was there first; just because I'm bigger and faster doesn't give me the right to push him out of the way. Therefore I told the tower I needed a right 360 for spacing.

As soon as I dipped the right wing, I thought, "Maybe this wasn't such a good idea." My Malibu wanted to drop out of the sky again. I nursed her around oh so carefully. As I cleared the airport boundary I dropped the gear, nose still down. Over the threshold I flared slightly as I added 10 degrees of flaps, and plop! the plane came down on the runway as if to let me know that in that configuration, it had no other choice. Every leading edge and every antennae was caked with well over an inch of solid ice. It literally took an ice pick from the restaurant to remove it.

When I asked FSS where the ice came from, I was told that the storm had developed very quickly, and I was the first to encounter it. While I was struggling to land, a sigmet on severe icing had gone out.

I spent the next day reviewing the experience with Piper. We came to the conclusion that all procedures were followed correctly, but the conditions were simply beyond the capabilities of the aircraft. I also realized that I should have declared a precautionary landing instead of making that very risky 360 prior to landing. A precautionary landing is not a declaration of an emergency, requires no paperwork, and is simply the most prudent thing to do when you're in a situation like mine.

✈ ✈ ✈

*The deicing boots weren't up to the challenge and the Malibu's performance was marginal at best. But the pilot elected to make things even tougher by executing a 360 for spacing. It was almost his own execution.*

*Flying across the Atlantic is always a challenge, but the pilot of a Twin Comanche found just how difficult it can be. After losing all navigation guidance he relied on a wind report from an Omega-equipped Baron. When finally located, he was way off course.*

# Glacier Greaser
### by Jeff Justis

My wife, Sally, and I were returning to the United States from Europe in my Twin Comanche. So far, the relatively primitive navigational aids along the North Atlantic route had served us adequately. Up there, VORs are a close-in luxury only, and intermittent loran reception pushed the ADF to the head of the class. A borrowed high-frequency radio supplemented our communications capability, but dead reckoning counted for as much as anything else.

Having stopped for the night in Iceland, we awoke early to Reykjavik's foggy skies. Weather was not available from Narssarssuaq, Greenland, our first destination airport, until nine a.m. local time, so we weren't in a hurry to get to the airplane. About four other pilots were receiving briefings at the same time, and I took my time figuring the winds from the 10,000-foot chart, adding 30 degrees of magnetic variation. Applying this to my true course indicated no significant correction and a 15-knot headwind component.

As I pushed full throttle to launch into the gray sky, I felt that everything was under control. That good feeling intensified as we broke out on top at 4,000 feet and climbed to 9,000. The ADF was steady, the VOR was working (at least close in to the station) and the loran was giving a reliable signal.

Communication remained good on the VHF radios, and when I reported over a fix 120 nm out, Keflavik Radar (in Iceland) remarked that I was eight miles south of track. Based on that, I made a five-degree correction to the north. I was in communication with a Baron that was equipped with Omega long-range navigation equipment. About one hour into the flight my loran became unreliable, but I was able to pick up two signals on the ADF, one at Kulusuk, Greenland, north of my position, and the other southwest of me at Point Christian on the southern tip of Greenland. I plotted my rough position based on those beacons, and it correlated with my previously known position and groundspeed of 150 knots.

Then I heard the Baron pilot say that his Omega indicated a southerly drift; he speculated that the wind had shifted to west-northwesterly. I added another five degrees northerly correction because I feared missing the tip of Greenland. I was still receiving Kulusuk and Point Christian. Once again my confidence grew when the landmass of Greenland came into view. As we got closer, it seemed that we were arriving sooner than I had expected on the basis of my previously known groundspeed. I remember trying to identify some prominent fix on the east coast. I had good communication with Sondre Stromfjord Radio remoted to Kulusuk and, as I had expected, we were cleared to overfly Greenland at 12,000 feet. The weather at Narssarssuaq was reported to be quite good, although we would require a letdown through broken cloud layers.

At this point I became concerned since I was not yet receiving Narssarssuaq's beacon on the ADF. We were now in instrument conditions at 12,000 feet, having passed the coastline. I should have either returned to the coast or deviated toward Kulusuk, which also has an airport. But I had been to Narssarssuaq three times now—I had even flown the tricky fjord approach—and I had never been to Kulusuk. I began changing frequencies on the ADF, hoping to pick up our destination's beacon.

Then one of a pilot's greatest fears became a reality—total loss of navigational guidance. I could no longer receive Kulusuk or Point Christian, and I was in solid instrument conditions. The airplane didn't seem to notice my dilemma, however, and there was no icing. I had about two hours of fuel remaining in the tanks.

I began calling Sondre Stromfjord Radio, hoping for a direction finding. They suggested I try Sob Story, a remote site high on the ice cap. No response. Then we lost communication with Sondre Stromfjord. One of the other pilots flying toward Narssarssuaq relayed my problem to Julianehab, another radio station near Narssarssuaq. They suggested I try 700 kHz, a strong commercial station in the area. Again the ADF was silent. I had mentally prepared to arrive at Narssarssuaq's beacon at 14:45 Zulu time, having departed Iceland at about 10:30. I had based my ETA on dead reckoning on passing the east coast of Greenland. I had also been flying for 45 minutes at 12,000 feet with no oxygen. Surely, the fateful decision I next made is a reflection of judgment distorted by hypoxia.

I looked at my ONC chart for elevations along what I thought was my route of flight and saw that 9,000 feet would clear everything in the area except a 9,000-foot peak farther north. Besides, on climb-out I had not entered the overcast until passing through 10,000 feet; 9,000 feet was also the en route altitude from Narssarssuaq to nearby Simiutaq. I was still talking with another pilot and advised him of my intentions. He relayed to me that Julianehab Radio advised against a descent.

So strongly had I rationalized my position that I had convinced myself I would soon break out of the overcast and find myself safe and sound over the west coast of Greenland. I therefore began a slow descent from 12,000 feet. The whiteness engulfed us completely. We continued through 10,000 feet toward 9,000.

Suddenly I felt a violent shaking of the left engine and I saw its propeller jerking. My first thought was of a catastrophic engine failure. I pushed everything to the firewall. The right engine roared and I think we flew again for a moment. Then I realized that the blades of the left prop were stopped dead and folded back toward the nacelle. Within seconds I was thrown back and forth in the seat and my head repeatedly hit the glareshield. Then we stopped, and for a moment I still couldn't see the ground—we were engulfed in white, but the softness of the clouds was not there. The right engine had stopped and its propeller was also strangely twisted.

"My God, what have I done?" I thought, and asked Sally if she was all right. Although she thought at first that her back was broken, in fact she was not seriously hurt. My nose was bleeding, but I was otherwise uninjured. I apologized for the crash and my wife agreed that, undoubtedly, it had been my worst landing to date. Both of us were able to laugh. We were alive and that was the important thing.

A passing jet relayed our Mayday call and within six hours a rescue helicopter was on the scene. We had crashed on Greenland's high ice cap, about 100 miles north of my intended course. Looking at a map showed me what had happened. We were at the center of the island in the middle of about 40,000 square miles of snow and ice. I had overcorrected for track error and the winds had been stronger than forecast out of the south. Apparently, the wind had not shifted to the northwest as I had thought based on the Baron pilot's report. In fact, with the wind's direction remaining southerly and its increase in velocity, I was compounding my error when I added northerly correction to my heading in fear of passing to the south of Greenland. Thus I had crossed the coast of Greenland at a point closer to Kulusuk (and to Iceland) than to Narssarssuaq, which accounted for my early arrival over the coast. My dead reckoning was therefore in error and I had let down onto the high ice cap rather than the lower ground to the south.

There was no undoing the accident and soon the old Twin Comanche that had served us so well would be entombed in the ice. As the rescue helicopter rose I looked back at the beautiful airplane on which I had worked for countless hours over the years. A wave of remorse spread over me as I abandoned it to the blowing snow.

*In some ways the pilot was lucky. He didn't know he was letting down onto Greenland's high ice cap. It came as a complete surprise that it was the ice cap and not a cloud he was trying to fly through.*

*In the next account, the pilot, again flying in the far north, knew what the conditions were when he took off on a medevac flight, but he felt the risks were worth it. That was before his vacuum pump decided to go south and left him out in the cold.*

## "Two Saves, Partial Panel"
### by Tim Olden

I was the standby medevac pilot for an Alaskan charter firm based six miles north of the Arctic Circle, so when the phone rang at 12:45 a.m. I knew it meant trouble. Snow had been falling for several hours and as I stood shivering, talking on the phone, I could barely see the lights not more than 400 feet from my window.

Our dispatcher in Fairbanks told me that drunken violence had broken out in a nearby village and there were several gunshot victims. One of the victims, before losing consciousness, had torched the home of his assailant in an apparent act of revenge. A number of burn patients required transport: all were in shock and would need emergency treatment in Fairbanks, 138 nautical miles to the south. They would certainly perish if treatment were delayed.

Just finding the village, 52 miles west of us, would be difficult, but the other two pilots and I agreed that we were duty-bound to give it a try. The landing would be on a snow-packed runway using snowmobile headlamps for runway lights—and our ski-equipped Cessna 185 was down for maintenance.

I took off first with a state trooper (also a qualified EMT) on board. Two other Cessnas followed, with our village nurse on board the third. If I didn't make it, the others would turn back. The local Air Force radar station vectored me to the village, as we had previously practiced in the event of such an emergency. I broke out of the clouds at 400 feet agl, two miles south-southeast of the village's primitive runway. The landing conditions deteriorated so much as I approached the ground that I had actually started to go around when the Cessna slammed onto the ground and skidded to a stop.

The second Cessna landed with a dramatic cloud of snow thrown upward, as did the last aircraft. Our patients were not in good shape. After the third airplane landed, the nurse set about readying them for the trip. My stomach rolled with the initial shock of seeing those who were burned. The trip would be hard on them. Soon, IV bottles were hanging from the Cessna's ceiling and oxygen bottles were in place. We secured the patients and were ready to depart.

With the yoke in my lap once we started to roll, the airplane pointed its nose skyward and eventually staggered into the air, tentatively at first, then in earnest. From here on, I thought, it's going to be a piece of cake. Just an hour or so of solid instruments, but the worst part of this flight is over.

About 15 minutes into our climb, having just picked up our clearance, both the artificial horizon and the directional gyro rolled in opposite directions. What a time to lose the vacuum pump. I've never liked partial-panel flying, and this was for keeps.

Covering the offending instruments so that I would not inadvertently rely on one or both of them, I began to think about how I would make the approach into Fairbanks. I told center that I would need no-gyro vectors en route and for the approach. In an attempt to keep myself alert, I made almost continuous calculations of our position, ETA, fuel burn—anything to keep the mind active. I didn't have long to wait for more trouble to keep me busy.

Something began to smell hot, like solder, or smoking insulation. I hoped that my imagination was working overtime, but the trooper noticed it, too. The cockpit was becoming smoky enough to make my eyes water. "Quick," I said, "turn the oxygen off. We're on fire."

"Fairbanks Lifeguard One, climbing 10,000 for 12,000. We've got a problem." Transponder to 7700. Fort Yukon AFS would have our radar position and could advise Fairbanks if I could not make radio contact on this one-time chance.

"Lifeguard One, Fairbanks."

"Fairbanks we have an electrical fire. Have to shut things down here. Estimate our position 81 miles northwest of Fairbanks, expect Fairbanks in 38 minutes. We will have to make a descent to visual conditions after passing Fairbanks. No gyros."

"Lifeguard One, cleared to cruise one-two thousand. How much fuel, and how many souls on board?"

"Fairbanks, we've got two-point-five hours plus, and four

souls. Gotta go—see y'all later."

We continued the climb through 11,000 feet, and broke out of the clouds on our way up to 12,000. At least for a while, we would be on top. The smell of burning insulation slowly faded. We had adequate fuel, good batteries in the flashlight and fresh extras in my flight case, and good weather—to the extent that we were not encountering freezing rain or icing, nor was it forecast. We were currently on top, but for how long?

Remembering the invaluable lessons learned on my instrument check ride, during which my examiner spent close to an hour demystifying the riddles of lead and lag on the magnetic compass, I flew on toward Fairbanks using the instruments I had left.

"These guys need the oxygen badly," the trooper said. "They're beginning to look anoxic. Can we risk it?"

I swept the cockpit with my light. The air was fairly clear.

"Yeah, go ahead."

Our patients' conditions dictated that we land as soon as possible. Bone-tired with fatigue, I felt willing to do anything to get down to the ground. Emergencies, though, are like wrestling with a gorilla: you don't stop when you're tired—you stop when the gorilla is tired. According to my notes, it was time to begin our descent, or should I add five minutes? Adding the time would put me farther from ridgetops close to the airport. Wanting to err on the side of conservatism, I added the time.

The trooper poked his head up front. "How much longer?"

"About 15 minutes." An answer based more on hope than fact. "How are the O.K. Corral boys doing?"

"Worse."

As we started down, workload and the late hour were combining to take their toll on me. I could not afford mistakes at this point, but I was desperately tired. I asked the trooper to help by holding the flashlight, and to call out altitudes.

"Five thousand." The turbulence was making heading control difficult. With the compass swinging from side to side, it was necessary to split the difference to steer our course.

"Four thousand."

Equally difficult was the instrument scan: trying to read the dimly lit instruments and compass in the bouncing airplane was straining my already overworked and tired eyes.

"Three thousand."

"Call out the altitude in 500-foot increments."

"Twenty-five hundred."

"Two thousand." If we were going to hit something, it would be soon.

"Fifteen hundred."

"One thousand." So far, so good.

"Call the altitude in 100-foot increments."

"Nine hundred."

"Eight hundred."

Ground contact. But, where are we?

"See any lights?" Staying on the instruments, I leveled off.

"Yes. Behind us," shouted the trooper.

We were out of the clouds, but it was snowing and visibility was poor. Looking around, I could see a string of lights, probably early-morning auto traffic, at about our seven o'clock position. We were a bare 400 feet above the ground, and to lose visual reference at this point could be fatal. I turned back toward the lights. I could see only the auto lights heading to the west—commuters on the way into Fairbanks, I hoped. I turned west in an attempt to find the airport.

I caught a dim flash of light to our right—was it an airport beacon? The layout of highways on the ground seemed right for Fairbanks International. In just a few seconds, as if the controllers in the tower had read my mind, the runway lights were turned up to maximum intensity. A bright, emerald-green point of light appeared at midfield. It was singularly the most beautiful thing that I've seen in my flying career. We were cleared to land.

There were no headlines in the local newspapers about the medevac, or the problems that plagued the flight. That's what pilots get paid for, I guess. My logbook contained a simple entry for the flight, "Medevac, two patients—two saves; partial panel," which doesn't do much justice to overcoming the trials endured. The payoff was that our patients survived.

✈ ✈ ✈

*Partial panel practice and a thorough understanding of the wet compass helped the medevac pilot keep his wings straight and his head up. In the next account the pilot knew his vacuum had failed, elected not to rely on a backup system, and chose to fly partial panel. Fooled into following the inoperative DG, the pilot drifted off course. Correcting back to course began a frightening roller coaster ride.*

# Vacuum Failure

### by Jonathan Apfelbaum

I am a young private pilot with 150 hours in my logbook. In the fall of 1984 I had had my certificate for two-and-a-half months. My father, also a private pilot, had more than 1,200 hours, an instrument rating and many hours flying with the Air National Guard as a forward air controller. The airplane we were flying was a Mooney 201 in which my father had hundreds of hours and I had a lot of experience but little time logged.

I was a senior in high school and looking at colleges. We decided to fly to North Carolina to see Duke University and fly home to New York that same evening.

After my interview at the university, we had a quick dinner and went back to the airport as the sun was setting. During the run-up I was careful to make sure that the vacuum gauge indicated in the green.

As we climbed into the night, I watched the lights of the town spin slowly around our left wing as we turned north, then disappear when we entered the clouds. We climbed to 11,000 feet between two cloud layers that obscured all outside light. With the cockpit lights lowered, only the DME told us that we were moving at all, and our senses called it a liar. We seemed to be hanging there in space.

I was flying from the right seat and Dad was talking to Washington Center when I noticed the first sign of trouble. The attitude indicator was sinking, but a quick check of the other instruments confirmed that we were straight-and-level.

Playing a flashlight beam around the cockpit, we checked each item on the checklist. The gyro itself wasn't the problem because the unit on the left side was also out of commission. The beam started across to my side to check the circuit breakers but stopped halfway on the vacuum gauge. Zero. No vacuum. The pump had failed, rendering useless our attitude and directional gyros.

Vacuum failures are not uncommon, and because we do a lot of IFR flying we have a backup system. However, the backup runs off the pressure differential between the manifold and the outside air, and at 11,000 feet the outside pressure was too low to run the gyros without a power reduction.

We decided to leave the power setting where it was and reduce it when we started down, reasoning that we wouldn't need the attitude gyro until then and we could fly the magnetic compass, airspeed indicator and turn-and-bank. It seemed safe. There was no foreseeable condition in which we would need the gyros until later and we would get home sooner if we didn't reduce power.

We covered the attitude indicator but decided just to ignore the directional gyro. But by habit, I followed the DG. It was taking so long to wind down that I suppose I assumed enough vacuum pressure was being produced by the backup to keep it running. I was wrong. Then Washington called a course correction to us. We were deviating from our flight path and should turn to 030 degrees. I entered a standard-rate turn and watched the DG. It showed 020 degrees and wasn't moving. Confused and disoriented, I leveled out and looked at the magnetic compass. We were going south—definitely not the way home.

Dad took over and turned hard right. Too hard. At night with no attitude indicator, we rolled upside-down and went into a steep spiral. The Gs built and we couldn't tell which way was up. Dad chopped the power to let the gyros spin up and tore the cover off the attitude indicator. The VSI was pegged, and the altimeter was unwinding fast.

At first there was sheer terror—like going up and over the first hill of a roller coaster. But after this sudden burst of fear came a calm. I relaxed. So this was how it happened, caught in a situation of our own making and now, beyond our control, we would die. I wondered how Mom would take the news. We would probably see the lights of civilization as we broke out of the clouds and watch them grow until we slammed into the ground. I thought of friends and places I had enjoyed and wasn't scared of dying, just curious and a little sad.

In retrospect, it was silly to think that way. It took only 20 or 30 seconds for the gyros to spin up, and at a descent rate of about 2,500 fpm we had four minutes before hitting the ground. The cloud base was at 5,000 feet, ample room to orient ourselves visually and recover.

At 7,500 feet, the gyros came alive and Dad reversed the downward spiral. We climbed back to 11,000 feet and rolled out on our proper heading. For the rest of the flight we kept the power reduced and flew with gyros. It took an hour longer to get home than the trip down that morning, but I personally savored every second of those 60 minutes.

✈ ✈ ✈

*The decision to fly without the back-up vacuum system turned out not to have been a wise one. But we always see better with hindsight. The pilot of a Mooney looked back on his night flight and realized just how close he'd come.*

# Low Profile
### *by John Morzenti*

It was one of those warm and humid late-June nights in Philadelphia, typical of the last days of a pleasant Bermuda high. Winds were southwesterly, and twilight would come a little after nine. Good, I thought as I drove through the gate at Wings Field outside of Philadelphia. I'd have a bit of a tailwind and would be home only slightly after dark. Home was Greenwood Lake Airport in northern New Jersey, 73 nm northeast, or about a 35-minute flight in the Mooney M20E.

It had been a long and tiring day. My mind was racing with a thousand details, few of which related to flying. As I belted in and began a cursory cockpit check, I reflected that while it was VFR in the haze, it was probably marginal at best. However, I'd flown the route many times and, frankly, I didn't feel like being sent hither and yon by New York ATC.

So VFR it was. Giving a final cockpit check, I set the altimeter and began taxiing as the oil temperature came off the peg. Once I was aloft the haze seemed no worse than it had appeared from the ground, and I had a good five miles visibility downsun. Pleased with my decision to go VFR, I took up a heading to Greenwood Lake.

The sun soon went down and ground lights began appearing through the murk. While the visibility was okay from the standpoint of seeing and avoiding other aircraft, it was getting a bit difficult to navigate visually. To make things easier, I decided to fly direct to Sparta VOR and make a modified VOR-A approach to the airport. This involved just a slight kink in my routing but would ensure finding the airport without difficulty.

Monitoring New York Approach to keep posted on any IFR traffic, I passed over Sparta and began my descent to the published MDA, 1,980 feet, turning to intercept the 079-degree radial. The only light at this point was the orange glow to the west, out the left cockpit window. The ground below, uninhabited reservoir property, was completely dark. Leveling off 2,000 feet, I had the vague sensation of large shapes hurtling by close below. Something seemed to flash by the left cockpit window—above my head, in fact. Then I was over the valley and finally entering the downwind leg for Runway 24 at Greenwood Lake.

I remember feeling strangely disoriented by the appearance of the field and the position of the landing runway. I seemed to be far too low as I turned base, then final, dragging in on the engine for an uneventful, if graceless, landing. As I taxied in to the ramp I puzzled over how I ended up at what appeared to be several hundred feet low on the downwind leg.

Just before shutting off the panel lights, a glance at the altimeter told me why. It read nearly 1,300 feet. Field elevation at Greenwood Lake is 792 feet. What was going on here? I distinctly remembered setting the altimeter to field elevation at Wings. Surely there couldn't have been that much barometric difference between Philadelphia and New Jersey in the big high-pressure system we were sitting in.

A little simple arithmetic told the tale pretty succinctly. I had set the altimeter all right. But I had set it for Greenwood Lake's field elevation, not Wings'. And because I didn't really expect any significant change in barometric pressure along the route of flight, I didn't bother monitoring ATC's altimeter updates to IFR traffic in the area. In short, I was nearly 500 feet lower than my altimeter told me I was, quite enough to make me a grim statistic on most nonprecision approaches.

As I later reconstructed, those vague shapes rushing by underneath were the rocks and outcroppings that make up the granite ridgeline two miles southwest of the airport. The object that seemed to flash by above me was a small ranger tower.

I learned a couple of lessons from this enlightening episode, aside from those that relate directly to proper altimeter usage. The first is that if I'm tired and distracted, I should either stay put or find some other way to get home. That goes double if the flight is to be in the dark or IMC. The second is that monotonous, back-and-forth commutation flying can breed a potentially lethal complacency.

And last, as my airline-captain father-in-law once told me, don't mix VFR and IFR procedures. I never fully understood what he was trying to tell me until that night.

✈ ✈ ✈

*The Mooney pilot set his altimeter wrong. It was that simple. The Piper Saratoga pilot who related the next account was bugged by strange readings on his airspeed indicator during an instrument approach. Luckily he broke out before he broke up.*

# Bugged Pitot

## by James Holloway

It was a beautiful August morning at the Nut Tree Airport—not a cloud was in the sky and the California sun was only beginning to heat up the Sacramento Valley. It was the kind of day that lulls you into forgetting that the whole world is not always VFR. A friend was arriving at Oakland Airport and I was all set to pick him up in my "new" Saratoga—with only 30 hours together, the Piper and I were still new to each other.

Shortly after leveling off I saw the weather over the San Francisco Bay. The entire area looked solid enough to land on. Oakland ATIS reported 1,400-foot tops with a 400-foot ceiling, and the glideslope was inoperative. While considering leaving my friend to the mercy of the rental-car counter, I groped for my approach plates and noted that the VOR/DME minimums were the same as the ceiling.

Bay Approach cleared me for the VOR/DME approach, and I began to test my slightly rusty skills. It couldn't be all that bad. The layer was only 1,000 feet thick, so I'd be in the clear in a few seconds.

On my second or third scan after entering the clouds, I noted that the airspeed was dropping and I eased in a little power. On the next scan, I noted that the airspeed was down to 65 knots. I felt a little rush of adrenalin and the cross-checks speeded up—attitude 10 degrees nose down; wings level; rate of descent 400 feet per minute; airspeed 55 knots . . . 55 knots?

Just a bit shy of panic, I applied full power and felt the reassuring kick of the turbo. I milked the flaps and hit every rocker switch on the left panel, knowing that one of them was the pitot heat. I'd never heard of pitot ice at 50° F., but what else could it be? As I fumbled with power, gear, flaps and pitot heat, my airspeed kept decreasing—I was falling out of the sky.

I knew that I needed airspeed so I applied forward pressure on the yoke. Although the attitude indicator was at 30 degrees nose down, I was still losing airspeed. At the same time that the controller called me with a low-level alert, I broke out at 400 feet. I advised the tower that I was having problems and noticed that my airspeed read 140 knots. Times like that make you doubt your sanity.

After landing I had the pitot system checked and everything appeared normal. By the time I picked up my friend, the fog had lifted and we flew home VFR with no further problems.

I spent the next two nights lying awake trying to determine what went wrong. Was I rustier than I realized? Was I subconsciously pulling on the yoke when I thought I was pushing? Was I going off the deep end?

After talking it over with my instructor, we decided to recreate the situation. We made several hood approaches at another field under the same configurations as at Oakland, which only proved that I was indeed rusty. But the airplane performed normally. That night we talked it over again and decided to recreate the approach at Oakland. The next morning, the ceiling at Oakland was 700 feet. I set up everything as it had been on the day of my ordeal.

As we entered the soup my instructor told me that I could take off the hood—this was for real. By the time I had removed the hood and rechecked the instruments, the airspeed indication had dropped to 65 knots from 100. I pushed the yoke forward and broke out at 700 feet and 50 knots. Within seconds, the airspeed read 145 knots.

We decided to try the approach again with the pitot heat on and the RNAV set up to read groundspeed. Again, everything was normal until we entered the clouds. Then the bottom fell out of the airspeed, but the groundspeed remained steady. I continued the approach based on groundspeed until we were VFR when, within seconds, the airspeed indication came up to normal.

On the way home we tried several times to recreate the situation under VFR conditions but to no avail. It was as if the Saratoga didn't want me flying in weather. Early the next morning several friends gathered at the hangar to witness an autopsy of the pitot static system. The lines were opened and air blown through. Neither moss, mildew nor any gremlins came scurrying out.

Next, the pitot head was removed and, when we turned it upside down, a small ball that looked like a check valve rolled back and forth in the lower port. A deft jab with a piece of safety wire removed the culprit. A small insect known as a leaf roller had set up housekeeping who knows how long ago. The leaf roller's nest was just the right size to allow enough air passage for normal testing and operation, but when the least amount of moisture collected, it would swell and block the passage of air.

Since this incident, I regularly inspect the underside of the pitot tube. It's difficult, but it's worth the effort.

# Other Bestsellers of Related Interest

**MORE I Learned About Flying From That**
*Editors of FLYING® magazine*
This second volume contains more stories from FLYING magazine's most popular column. These first-hand accounts submitted by active pilots not only furnish exciting reading but offer valuable tips that fliers can use in the cockpit. Entertaining and informative for pilots and armchair aviators everywhere.
**0-07-155418-1   $14.95 Paper**

**ABCs of Safe Flying, 3rd Edition**
*David Frazier*
This book gives you a wealth of flight safety information in a fun-to-read format. The author's anecdotal episodes plus NTSB accident reports lend both humor and sobering reality to the text. Detailed photographs, maps, and illustrations ensure that you understand key concepts and techniques. If you want to make sure you have the right skills each time you fly, this book is your one-stop source.
**0-07-157718-1   $15.95 Paper**
**0-07-157719-X   $22.95 Hard**

**Handling In-Flight Emergencies**
*Jerry Eichenberger*
Almost all of the emergencies you'll face as a pilot are manageable—if you have the correct training and mental attitude. With this guide, get the most out of preflight preparation, be more aware of changing flight conditions, and learn the correct response for the most common cockpit emergencies you're likely to face. After reading this book, you'll be better equipped to handle any emergencies that come your way.
**0-07-015093-1      $21.95 Paper**
**0-07-015092-3      $34.95 Hard**

## Be A Better Pilot: Making the Right Decisions
*Paul A. Craig*
Why do good pilots sometimes make bad decisions? This book takes an in-depth look at the ways pilots make important pre-flight and in-flight decisions. It dispels the myths surrounding the pilot personality, provides straightforward solutions to poor decision traits that affect the way they approach situations.
**0-07-157664-9   $17.95 Paper**

## Design for Safety, 2nd Edition
*David Thurston*
Thurston reviews the principal causes of general aviation accidents and explains why some accidents are really not so much the result of pilot error as they are the fault of poor design, poor flight instruction technique, inadequate air traffic control and communication procedures, and questionable airport location. Fortunately, these factors are subject to correction and improvement and this book recommends steps to effect needed changes.
**0-07-064560-4   $27.95 Hard**

## Van Sickle's Modern Airmanship, 7th Edition
*Ed. John F. Welch*
By far the most comprehensive reference on essential flight principles, techniques, and performance standards, this book has been used by pilots for more than 30 years. This updated and expanded edition reflects the latest in airplane and aerospace structural designs, engines and instruments, aeromedicine, satellite-based navigation, helicopters and gyrocopters, and gliders and sailplanes.
**0-07-069184-3   $44.95 Hard**

## Aviator's Guide to Navigation, 2nd Edition
*Donald J. Clausing*
Navigate safely around the skies using the expert advice you'll find here. This second edition gives you updated information on Loran-C, inertial navigation systems (INS), NAVSTAR/GPS, electronic flight information systems (EFIS), and Omega/VLF navigation systems. It's designed to help you understand and use the system that will best satisfy your navigation needs—and stay within your budget.
**0-07-011291-6   $19.95 Paper**

## Avoiding Common Pilot Errors: An Air Traffic Controller's View

*John Stewart*

This essential reference—written from the controller's perspective—interprets the mistakes pilots often make when operating in controlled airspace. It cites situations frequently encountered by controllers that show how improper training, lack of preflight preparation, poor communications skills, and confusing regulations can lead to pilot mistakes.

**0-07-155395-9     $18.95 Paper**

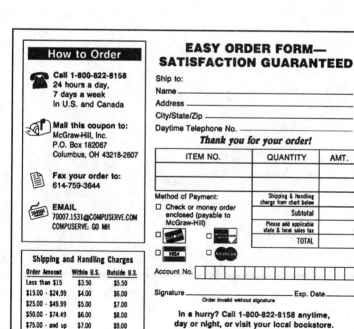